104 EASY Projects for the Electronics Gadgeteer

By Robert M. Brown

FIRST EDITION

FIRST PRINTING—APRIL 1970
SECOND PRINTING—MARCH 1971
THIRD PRINTING—AUGUST 1972

Printed in the United States
of America

Hardbound Edition: International Standard Book No: 0-8306-0524-X
Paperbound Edition: International Standard Book No: 0-8306-9524-9

Library of Congress Card Number: 72-114714

Contents

INTRODUCTION

This could very well be a publishing first—a book devoted to the premise that most of the real fun in electronic tinkering is derived from scrounging and swiping, trial and error, and good, old-fashioned cheap-skating. You'll be surprised at the useful and fun things you can put together with a little imagination and available components. You'll be even more surprised when you realize what can be done without tubes or transistors! The purpose of this book is to give you some food for thought along these lines and to clue you in on some tricks of the trade in getting the most for the least.

THE CIRCUITS

The circuits presented in this collection have been selected and designed to use as many garden-variety common electronics parts as possible. The components are generally found in old TV sets and other household electronics instruments. We have used few actual components in each circuit so that even if you went out and bought everything new, it shouldn't run you more than $4 or $5!

But don't get the idea that these circuits are straight from some company's design labs, representing hundreds and thousands of dollars worth of engineering and research. For the most part they are a collection of gizmos which were whipped together by the authors out of sheer necessity to fill a particular application. You will find that many of our circuits may be combined with each other to form an entirely new device.

Those of you with some previous experience in electronics will undoubtedly look at some of these circuits and say, "Why doesn't this particular circuit have two or three more components to give it double the efficiency," or "double the power," or "to enable it to perform such-and-such extra task." Well, we built 'em as we needed 'em and make no claims that any or all achieve any particular plateau of sophistication. By all means, if you can modify or otherwise improve them you are invited to do so. (Be sure to let us know the results.)

Since few of these circuits are of "critical design," it should be pointed out that the components shown in the parts lists represent the ones we used when building the original working model. You should, therefore, not panic when you come across a part having a specific value which you can't readily locate. If we show, say, a 47K resistor, the circuit will undoubtedly work just as well with any number of other resistors having values in the same general region; they might very well work with a resistor having a greatly different value. Any critical components are indicated "Do not substitute" somewhere in the text; otherwise, you're welcome to be your own Edison.

OBTAINING PARTS

The best source of parts for building these circuits is right around your own home. Most of us have basements, attics, or closets which are chock full of junky pieces of electronics gear which will probably never again be used; this includes TV sets, AM and FM radios, audio gear, tape machines, record players, etc. These are goldmines of components and should be attacked vigorously with wirecutters and screwdrivers. Take everything, cannibalize them right down to the bare metal chassis. Drill out the rivets holding down the sockets, pry up the IF cans, take out the terminal wiring strips, even the knobs.

All resistors and capacitors should be cut loose, leaving the leads as long as possible. When removing wired-in diodes, transistors, and neon bulbs (such as the little NE-2s), try to unsolder them, but be sure to place a pair of tweezers on each lead while unsoldering so that the heat will not travel up the lead and damage the component. (This is a good trick to remember when putting them into a circuit, too.)

If you don't have any equipment which you feel like butchering, then go down to your nearest radio-TV service shop. They undoubtedly have a whole slew of such sets gathering dust with which they will gladly part for a dollar or two.

BULBS AND DIODES

The first thing you, as an experimenter par-excellence, must get is a thorough industrial electronics catalog (we use Lafayette, Allied, and Cramer Electronics; the address for Cramer's 528-page monstrosity is 320 Needham St., Newton, Mass. 02164) which will tell you which pilot lights, neon bulbs, SCRs, selenium rectifiers, and crystal diodes you can switch around and substitute in any given circuit. Additionally, there is a substitution guide in the back of this book which should come in extremely handy when you're caught without a 1N38B.

In the event you simply can't locate the desired component or an acceptable substitution, then you are going to have to buy one. Before you dash out and purchase a name-brand bottle, check out the dealers handling surplus electronics parts.

PARTS, IN GENERAL

Until you build up a fully-stocked junk parts box (it takes a few years), you may find that a few parts will have to be bought that would normally be considered available, such as SPST toggle switches, 3-amp 117v AC fuses, hookup wire, resistors, potentiometers, capacitors, etc. For your information in locating the needed part at the best possible price, we are including a list of the major low-cost components suppliers. This includes parts distributors, surplus dealers, even major mail-order suppliers who can offer low prices because of the large volume of business they do. Our suggestion is for you to invest a dollar or so and dash off a postcard to each and ask to be placed on their catalog mailing list. This will enable you to shop with ease for the very best deal.

CHASSIS

Most of the projects in this book were constructed in and on upside-down baking tins, a few were built in tuna cans, cigar boxes, on Vector Boards, and even 3 x 5 file card boxes, although we have in some instances suggested appropriate chassis dimensions as a general guideline. Of course, you can buy a ready-made chassis if you want.

CIRCUIT HINTS

You'll notice that each schematic uses a more or less standard format for listing the parts values. Resistances are in ohms and unless otherwise indicated all resistors are rated at 1/2 watt. Resistors marked "K" are in thousands of ohms, and those marked "meg" are in megohms. A certain value resistor followed by "pot" signifies that this is a potentiometer or variable resistor.

All capacitive values are either in microfarads, signified as "mfd," or picofarads, "pfd." In case you may be accustomed to some other style, rest assured that "mfd" is the very same as "mf" or "uf," and "pfd" is the same as "mmf." In the diagrams you'll notice that one capacitor plate on each component is curved, a fact you can usually ignore, although on electrolytic types this tells you that the curved end is the eventually-grounded or "minus" side. The straight plate is the "plus" or positive side.

Electrolytic capacitors, which are frequently used in these circuits, have a voltage rating. An example of this might be "16 mfd, 150 WVDC electrolytic." This means that the capacitor, which is generally much larger physically than other types, has a mazimum working voltage DC of 150. The voltage rating depends entirely on the DC (battery) voltage. Always use the next highest rating than the voltage used in the circuit if you can't match the requirements suggested in the parts list.

Always watch electrolytic polarity carefully. If you're not sure which end ("plus" or "minus") is which, look for the curved plate—you're key to the minus sign—and you're in the ball park. Also watch polarity on all diodes!

Note: On circuits employing AC in direct contact with electrolytic capacitors, double the next highest rating (WVDC) of the capacitor if you can't meet what we've suggested. This is a safety factor that may save you a lot of aggravation later on.

1

HOMEBREW LIGHT METER

A light meter, with proper calibration, has a number of useful applications for measuring illumination levels in the office, home, and outdoors, as well as indoors. Additionally, it can be effectively employed with printing boxes and photographic enlargers in the darkroom.

This remarkably inexpensive gadget can be put together in an hour, yet provides all the important features of commercial light meters—and at a fraction of their cost. While we have suggested your using an International Rectifier Corp. Type B5M or DP-5 self-generating photocell, practically any high-output type can be applied just as satisfactorily. Your 0-1 DC milliammeter can be calibrated in footcandles for each range of the switch by comparing your homebrew meter with a commercial type having similar calibration.

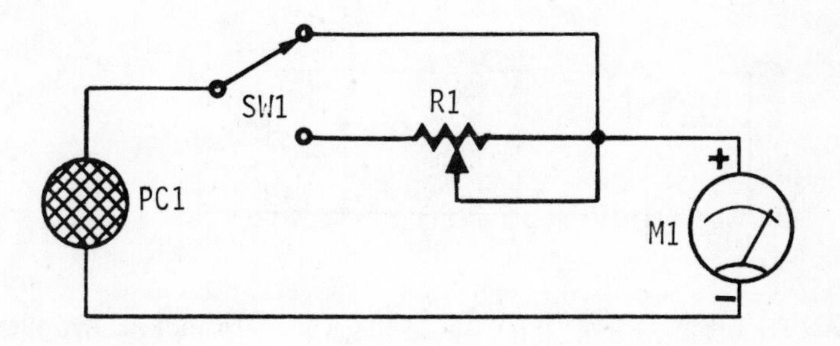

PARTS LIST

M1—0-1 DC milliammeter
PC1—Photocell. International Rectifier B5M or DP-5
R1—5K pot. Ohmite CU 5021
SW1—Low-power SPDT. Oak Type 189 lever switch

You can build this gadget into a surprisingly small metal box and, if you so desire, equip it with a handle for added versatility. (NOTE: With the components suggested here, 150 foot-candles will deflect a standard 55-ohm 0-1 DC milliammeter to full scale.)

2

TUBELESS/TRANSISTORLESS CODE PRACTICE MONITOR

Studying for your Novice Class amateur radio license? Or do you just like to experiment with the International Morse Code? Whatever your reason, here's a dandy CPO (short for code practice oscillator) that you can put together in a few minutes, using standard junkbox parts to provide a clean tone whenever you depress the key.

The neon bulb "blinks" each dash and dot at the same time you hear the tone in your earphone, thereby familiarizing you with not only the sound of International Morse, but also with what it "looks" like.

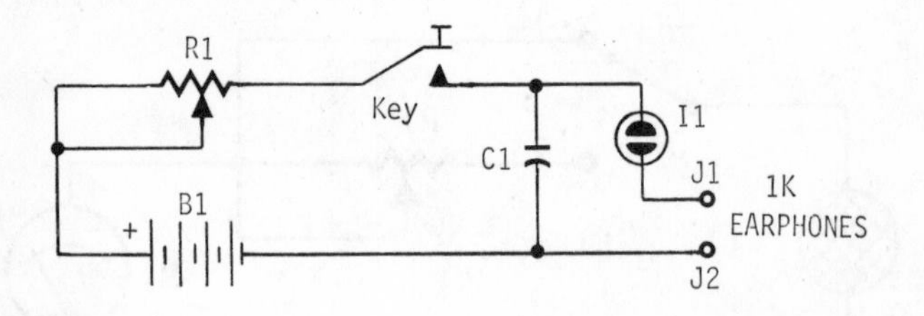

PARTS LIST	
B1—90v DC	J1, 2—Tip jacks. Amphenol 350-61001
C1—.0022 mfd	R1—1 meg pot. Ohmite CU 1052
I1—NE-2 neon bulb	

You can build this CPO in a pillbox (you can get several of these gadgets into an empty cigarette pack if you're handy with close, sub-miniature construction techniques). The battery can be anything from 90 to 125v DC. You may want to string five inexpensive 22 1/2v DC hearing-aid batteries in series for the required voltage, use a single Burgess-variety cell, or even a voltage quadrupler as shown in Project 102. NOTE: If you'd like a fixed tone, replace R1 with a fixed 407K resistor.

3

ELECTRIC COMBINATION LOCK

An electric combination lock can be very useful for a much-used door, such as the one into the workshop or garage. And magnetic-type door releases are quite inexpensive these days. A number of pushbutton switches can be connected in series in such a manner as to actuate the door release when all switches are closed.

Though simple in schematic form, the trick is in arranging the switches in a pattern so as to fool anyone not familiar with the "secret combination." It's all in how you mount them. You can twist around the switches so that what would normally ap-

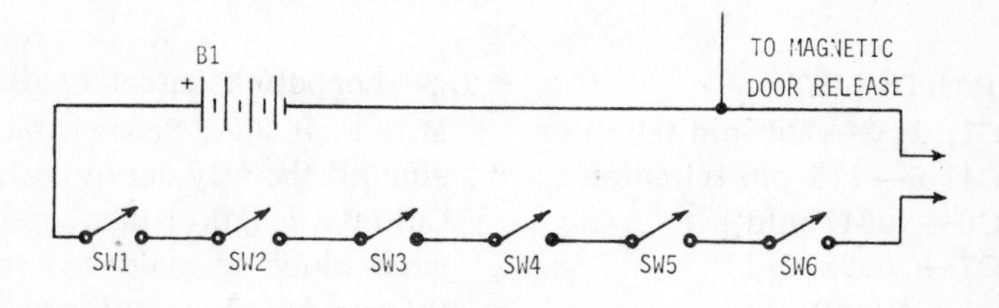

PARTS LIST
B1—DC as required for door release mechanism
SW1, 2, 3, 4, 5, 6—SPST pushbutton switches. Oak Type 175

pear to be an "on" position (to the right) is actually an "off" and vice versa. And you can mix up the arrangement, say, into two columns of three switches each. Whereas normal inclination would be to assume that the first switch in the left hand column is switch 1, it is actually switch 4, etc. Only you know the combination.

Word to the wise: Keep an extra "combination cribsheet" tucked away somewhere where you can find it in an emergency. Otherwise you may be stuck in the rain someday for an hour and a half, trying to remember the proper switching sequence!

4

SUPER CRYSTAL RADIO

Everyone at some time or another has built a crystal radio receiver, but we'll bet you've never tried this one! In operation this configuration is just one step short of being a high-quality fixed-tuned AM tuner, suitable for hookup to any good hi-fi amplifier.

This set is different, essentially, because it incorporates a switch, permitting you to "switch" to your favorite stations instead of slowly tuning them in. Its function is something like a pushbutton AM car radio—pre-set to those stations you listen to most of the time.

PARTS LIST

C1, 2, 3—500-pfd trimmer
C4, 5—175-pfd trimmer
C6—.0047 mfd
C7—.0022 mfd
D1—1N38B
J1, 2, 3, 4—Phone-tip jacks. Amphenol 350-61001
L1—5 turns of hookup wire wound around one end of L2
L2—Loopstick antenna coil. Miller #6300. Screw the slug all the way out of coil. Cut off a length of insulated wire which is soldered to the coil before installing.
PL1, 2, 3, 4—Phone-tip plugs. H.H. Smith 108
SW1—Double-pole, 5-position rotary switch

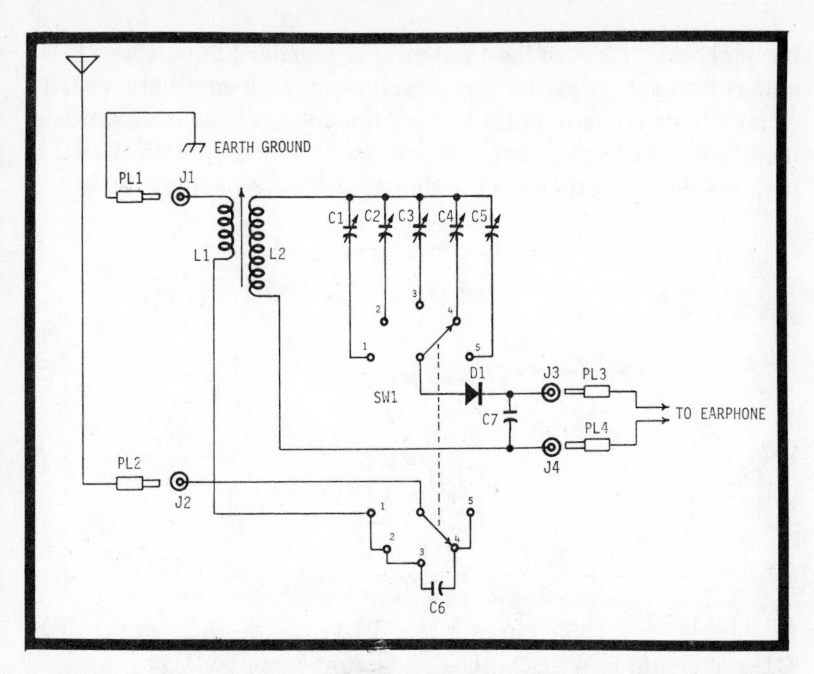

Use a good antenna and ground (an outdoor pipe driven at least four feet into the earth or a nearby cold water pipe), and a conventional magnetic earphone or headset rated at 2000-3000 ohms.

Presetting the trimmer capacitors is easy. Just switch to the first capacitor and adjust it until the station you desire is coming in loud and clear. Mark the station's call letters on your switchplate and then proceed to switch position 2. Follow this same procedure all the way through, and you're actually set for life. Unless your station goes off the air or changes operational frequency, you'll never again have to tune the radio.

5

SURPLUS RELAY LAMP FLASHER

Here's a unique lamp-flashing system you can put together for little or no cost. Nearly all the parts can be scrounged from the junkbox. Heart of the flasher is a standard 2000-ohm surplus relay such as can be found in numerous pieces of discarded Army/Navy control units and equipment such as hams often modify for their own use. If you can't come up with the relay, invest 48¢ and get one mail-order from a surplus dealer.

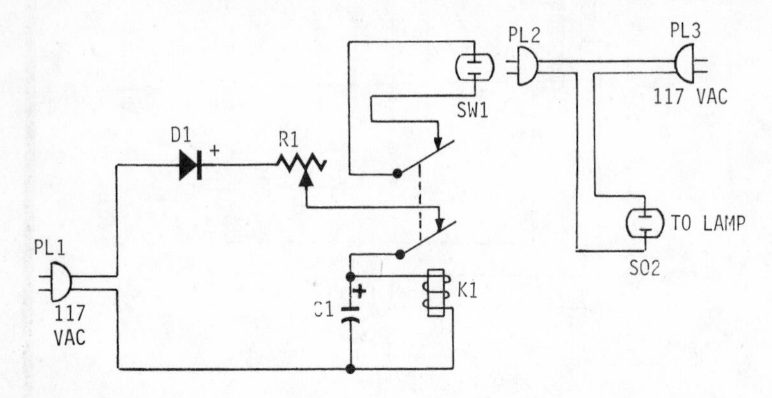

PARTS LIST
C1—500-mfd 50 WVDC electrolytic
D1—117v AC 35 ma selenium rectifier
K1—Surplus DPDT relay. See text.
PL1, 2, 3—AC wall plug. Amphenol 61-M11
R1—25K pot. Ohmite CU 2531
SO1, 2—AC socket receptacle. Amphenol 61-M1P-61F

Using the parts values shown, your flasher will cut in at the rate of approximately once each second, although you can regulate this to a great extent by varying the setting of R1.

If you don't have the rectifier handy, forget about 117v AC operation and simply hook any handy power pack (of up to 300 volts) across the input. Readjust R1 and you're in business!

Note: Key to guaranteed success is a fresh capacitor. If your flasher isn't up to par, replace C1.

6

SIGNAL GENERATOR

Here's another trick using a neon bulb that not too many people are aware of—a code-practice oscillator configuration that doubles as a great signal generator, providing you know how to use it.

Connect the minus side to the chassis of the unit under test and apply the positive probe to the stages of the set under study, starting with the output and working back. Try the speaker voice coil first, then the speaker transformer, the output tube (plate and grid), etc., until a point is found where you hear no oscillation in the speaker. At this point you've successfully isolated your problem and can proceed by checking the various components (tube, resistor, capacitor) for the culprit.

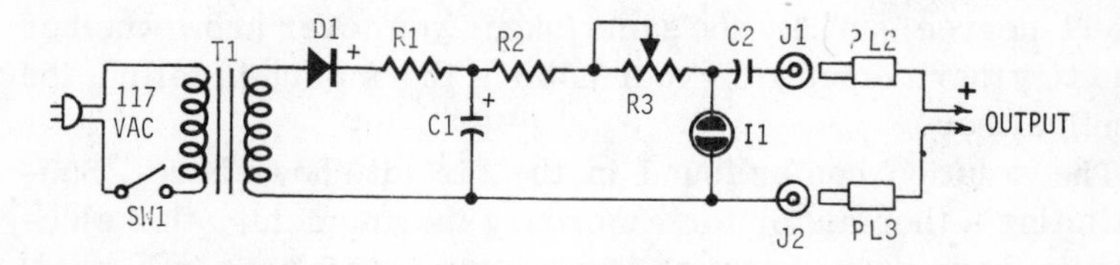

PARTS LIST

C1—16 mfd, 150 WVDC electrolytic
C2—.0022 mfd
D1—75 ma selenium rectifier
I1—NE-2 neon bulb
J1, 2—Standard tip jacks. Amphenol 350-61001
PL1—AC wall plug. Amphenol 61-M11
PL2, 3—Phone tips. H.H. Smith 108
R1—300 ohms, 1 watt. Ohmite "Little Devil"
R2—510K. Ohmite "Little Devil"
R3—1 meg pot. Ohmite CMU 1052
SW1—Single pushbutton AC power switch. Oak Type 175
T1—1:1 ratio isolation transformer

Incidentally, you can make a fierce siren out of this thing by adjusting R3 while you've got the output feeding into a high-wattage audio amplifier. A great gadget if your mobile CB rig has a provision for public address (PA switch)!

7

ELECTRONIC THERMOMETER

While conventional wall thermometers are inexpensive and handy, have you ever taken note of temperature readings on a group of identical types in a store? One five-and-dime we visited had 15 on a display rack, with only two agreeing as to the temperature! From the highest to the lowest there was an 11-degree gap! By the same token, you never know whether the tiny thermometer on your home's thermostat is telling the truth either.

The solution can be found in the circuit shown here. Substituting a thermistor for a mercury thermometer, this electronic device can measure temperature variations to a small fraction of a degree with surprising accuracy.

The trick is in calibrating your 0-1 DC milliammeter so it reads in degrees. Obviously, you will have to depend on a trustworthy thermometer for this, although if you take enough pains, you'll wind up with an electronic thermometer that'll

PARTS LIST

B1—30v DC
M1—0-1 DC milliammeter
R1—500K pot. Ohmite CMU 5041
R2—11K. Ohmite "Little Devil"
R3—5.1K. Ohmite "Little Devil"
R4—50K pot. Ohmite CMU 5031
SW1—Pushbutton. Oak Type 175
TH1—31D7 thermistor or equivalent

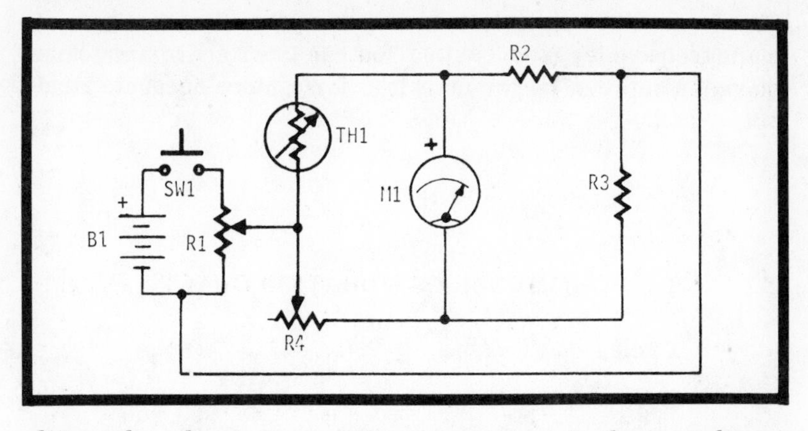

ultimately relegate your old mercury-type to the scrapheap. For this purpose you can either place a "read-out" cribsheet alongside the meter panel or tear into the meter itself and make a new face for it. Regardless, all you'll have to do is push SW1, and presto! Instant electronic temperature reading.

8

AUDIO FREQUENCY METER

With only two capacitors, two resistors, and a dual potentiometer you can make this easy-to-build audio frequency meter and measure any audio frequency between 50 and 10,000 cycles (Hz).

For convenience's sake, we suggest you build this into a sufficiently large Minibox to permit your installing a pointer dial knob with a large "swing" area on R4-R5. This is necessary because you'll want to calibrate R4-R5 to some detail in order to interpret the incoming audio. R1, your "fine tune" adjustment, should be mounted on the side of the box and more or less preset. You shouldn't have to adjust this very often.

For calibration purposes it is easiest to use a reliable audio oscillator and merely copy the proper frequency reading onto your R4-R5 dialplate. If you don't have an audio oscillator but do have an ohmmeter, however, you can hook the ohmmeter across either R4 or R5, take readings, and convert these to audio frequencies (see Table). You can later approximate frequencies between larger markings for a more accurate readout.

OHMMETER CALIBRATION CHART

Cycles	Ohms
50	320,000
100	160,000
300	53,000
500	32,000
800	20,000
1000	16,000
3000	5300
5000	3200
8000	2000
10,000	1600

Finer degrees of calibration can be determined by using proportionate figures between those shown above, although visual approximation from the calibrated dialplate is usually sufficient. Note: Use this method (of taking ohmmeter reading of either R4 or R5) only if good quality audio oscillator is not available.

To use your audio frequency meter, merely attach a pair of headphones across the output to monitor what's coming in. Once a signal is heard, adjust R4-R5 until you've successfully tuned it out. When no further tone signal is heard, the readout on your calibrated dialplate indicates the proper audio frequency. Simple?

PARTS LIST

C1, 2—.022 mfd
J1, 2, 3, 4—Phone-tip jacks. Amphenol 350-61001
PL1, 2, 3, 4—Phone - tip plug. H.H. Smith 108
R1—2K pot. IRC 11-110
R2—1.1K. Ohmite "Little Devil"
R3—2.1K. Ohmite "Little Devil"
R4, 5—1 meg pot. Ohmite CCU 1052

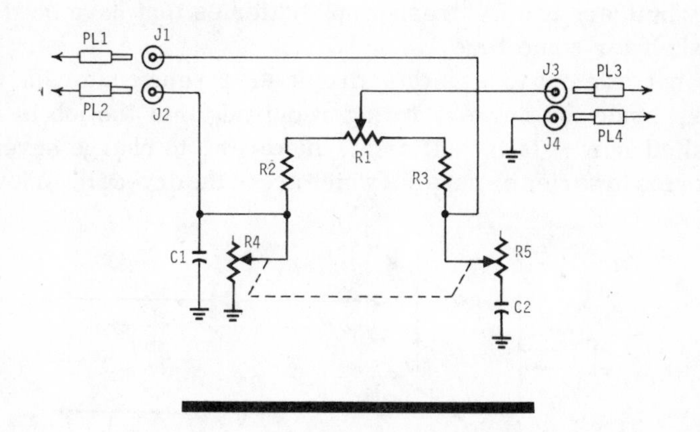

9

DRY-CELL REJUVENATOR

While there are many circuits for battery chargers, too few of them are capable of true cell rejuvenation. Often they will recharge some batteries temporarily, only enough so that they'll function briefly, but often the amount of applied current is insufficient to really give the cells a legitimate boost. This circuit, however, is designed for the truly "tough jobs"—the batteries that just can't be put back to use with ordinary charging methods.

Designed to handle even the truly heavy dry cells (up to the big 67 1/2-volt jobs), this charger supplies a healthy amount of current as well as voltage.

Notice that resistor R1 can be adjusted so that any number of resultant voltages can be provided. If you desire, substitute a heavy (10-20 watt) potentiometer for R1 and measure the resultant voltage with a voltmeter prior to placing the battery or batteries in for charge. Always make certain that the voltage you arrive at is always a bit higher than the battery's rating. For example, if you are charging a 9-volt TR battery, you can feed in about 12v DC (measured at no-load condition). With the values shown, the charger is ready to handle the 67 1/2-volt types.

Notice that the minus side of the rejuvenator is above ground. This must not be handled or touched, since a severe shock can result. It should be mentioned that this is the same circuit many builders use to "freshen up" batteries that have been on the shelf for some time.

Do not attempt to use this circuit as a rejuvenator on wet cells, since the current rating required to do the job is not supplied here. It is sufficient, however, to charge several batteries in series or parallel which are of the dry-cell variety.

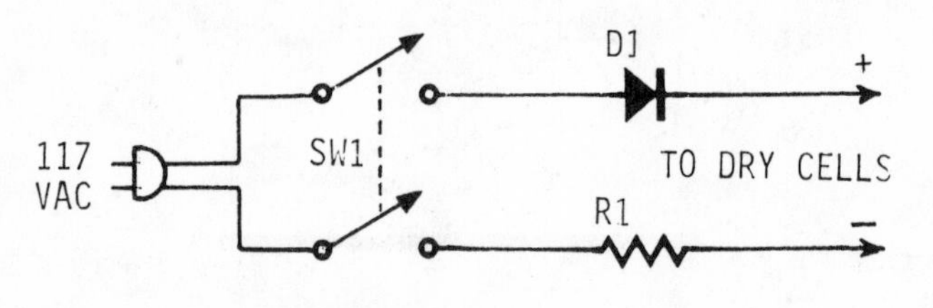

PARTS LIST

D1—75 ma selenium rectifier
R1—2.25K, 12 watts. Ohmite "Brown Devil" Type 1743
PL1—AC plug. Amphenol 61-M11
SW1—DPST. Oak Series 200

10

FLASHGUN TESTER

Nothing is so exasperating for a camerabug than to have one of those never-to-be-repeated scenes lost because of the failure of a flashbulb to fire. Yet this seems to happen to almost everyone at one time or another. Take heart, friends! For less than $1.50 (and nothing if your junkbox is a good provider) you can build this simple tester that will tell you ahead of time whether or not your flashbulbs will flash.

The secret of this project is the base of a used flashbulb which is remotely wired to the tester. Simply crack off the glass, leaving the socket intact, and hook this as shown in the diagram. Prior to taking your picture, slip the wired flashbulb base into the flashgun and cover the lens so as not to expose the film. Now close the shutter.

If the neon bulb flashes, so will your flashbulb. If nothing happens, you've found your trouble. Thoroughly clean the contacts and check the battery clips. Also check the condition of the batteries. Now try it again. Only when the neon lamp lights can you be assured against failure.

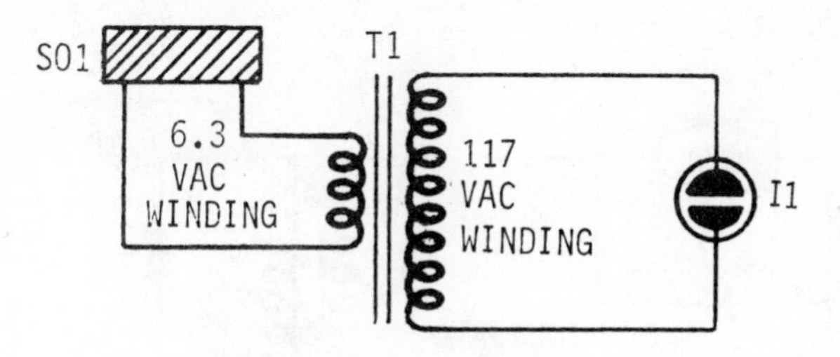

PARTS LIST

I1—NE-51
SO1—Old flashbulb base
T1—6.3v AC filament transformer. Triad F-14X

11

ULTRA-CHEAP SIGNAL GENERATOR

Looking for a reliable signal generator you can throw together inside of an hour from junkbox parts? Well, give this project some thought. The buzzer generator can be built into practically any container you like, has no critical components, and will do an admirable job of pumping buzzer-tone oscillations into your receiver.

This one has been designed to cover the 540-1700 kHz AM broadcast band, but you can modify the circuit by simply changing the values of L1-L2 and C2 to resonate at practically any frequency range that you might desire. Simply check the LC combinations with a good grid-dip meter prior to final installation, and you're in the ball park. Hint: To some extent you can halve the specifications shown every time you want to double the output frequency. For example, if you want to hit 2.40 MHz, you can use a 175-pfd variable for C2 and merely use half the windings called for in the parts list for L1-L2. A GDO should still be used for final pruning, however. To sweep to the correct frequency, just tune capacitor C2 until you hear the generator zero-beat with the signal you're listening to.

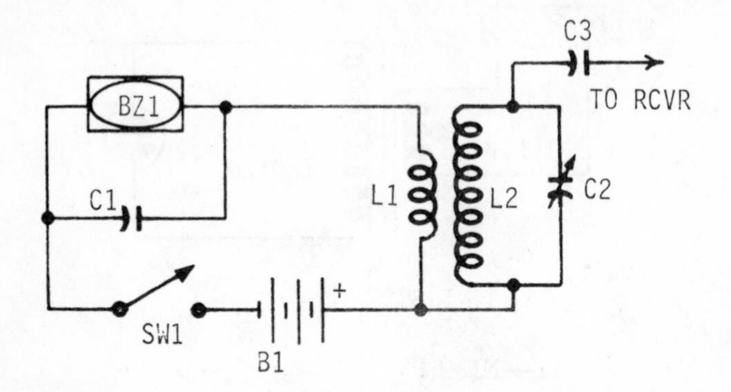

PARTS LIST

B1—3v DC
BZ1—3-volt buzzer
C1—.22 mfd
C2—360-pfd variable
C3—.022 mfd
L1, L2—Single Miller 20-A RF coil modified as follows: clip the fine wires leading to the short, adjustable sleeve and remove the entire sleeve from the form. Wind a new primary of 9 1/2 turns of hookup wire over the cold end of the coil. This is L1. Remainder of the coil serves as L2.
SW1—Single pushbutton power switch. Oak Type 175

12

TV SET RADIO RECEIVER

Few people are aware that their living-room TV set often comes with a phono input jack through which you can play a record player, FM tuner, or even a subminiature AM radio. Indeed this project is just that—a tiny yet tunable broadcast band receiver which is constantly "on."

Making use of one of the TV antenna twinlead wires for signal pickup, the small diode receiver couples directly to the phono input through a standard RCA phono plug so that you can remove the receiver whenever you like. If installation adversely affects TV reception, isolate the pickup wire by inserting a 15-pfd capacitor between the radio add-on and the antenna lead.

You'll find you can conserve space by employing a small mica trimmer capacitor instead of a conventional variable. Additionally, you'll probably find that you will want to listen to a

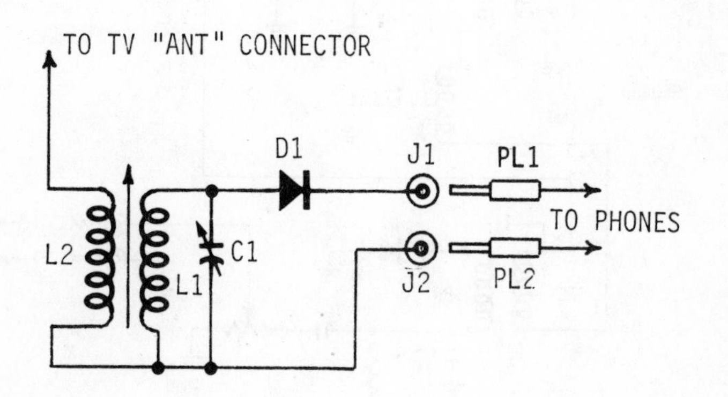

PARTS LIST

C1—365-pfd trimmer
D1—1N38B
J1, 2—Standard phone-tip jacks. Amphenol 350-61001
L1—Ferrite core. Miller 6300
L2—35 turns of #28 enameled wire wound over L1
PL1, 2—Phone-tip plugs. H. H. Smith 108

particular station when you switch the set into the "Phono" position, so you can simply preset the LC circuit to that station and leave it at that.

Suggestion: Many local stations are now offering "24-Hour-News" broadcasts. Why not tune into one of these stations? This way you'll always have a source for instant news. All you do is flick the switch!

13

HI-FI TUNER

Here's a selective AM tuner that will receive the full range of frequencies broadcast by local stations, yet cost you very little to build. Since it uses no tubes or transistors it requires no

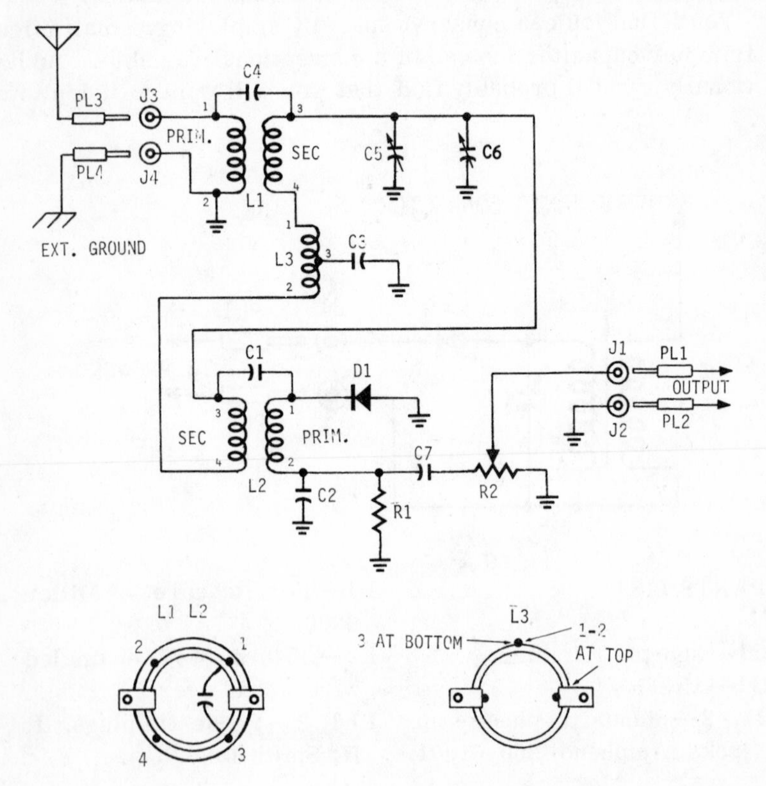

PARTS LIST

C1, 4—Included with L1 and L2
C2—2 pfd
C3—.002 mfd
C5, 6—365-pfd variable, double-ganged tuning capacitor
C7—.047 mfd
D1—1N38B
J1, 2, 3, 4—Standard phone-tip jacks. Amphenol 350-61001
L1, 2—Miller 242A (including C1, 4) TRF coils
L3—Miller EL-55 mutual coupling coil
PL1, 2, 3, 4—Phone-tip plugs. H.H. Smith 108
R1—110K. Ohmite "Little Devil"
R2—1 meg pot. Ohmite CMU 1052

power supply, yet it provides ample output across the entire AM band.

Aligning the receiver is quite simple. Connect a good antenna and ground to the unit and plug the output into your hi-fi amplifier. Incidentally, if you have a reliable preamp, you'll find that it will allow you to pull in the weak stations you wouldn't ordinarily hear. Adjust your tuning dial so that it reads at the extreme low edge of the band when the capacitor plates are fully meshed. Now tune in a strong station near the high end of the band (say, from 1300 to 1650 kHz). Does your dial read correctly? If not, adjust the small trimmer capacitors mounted on the sides of C5-C6 to shift the frequency of the station towards the proper point on the dial. Continued trial-and-error (adjusting both trimmers at each try) will allow you to relate your receiver to the most demanding commercial dial mechanism you can buy. If you don't require an elaborate tuning dial, you can set up your own calibration a lot simpler.

By the way, if you live in a rural area far from congested metropolitan broadcasting, you can add a 17-pfd capacitor across terminals 1 and 3 of coils L1 and L2 to bring in more signals. Although this boosts sensitivity immeasureably, some degree of loss of selectivity will result. In any case, simply add the 17-pfd capacitors without removing C1 and C4.

14

INEXPENSIVE ALARM

Here's a unique project that is bound to get you to work on time in the morning. Essentially, it consists of a photocell and an inexpensive Mallory Sonalert alarm hooked to a 25v DC source.

By running a pair of remote wires to a nearby window where sunlight is bound to strike in the AM, all you have to do is turn SW1 to the "on" position the night before. Next morning, prepare for a screeching howl that will continue blasting until you walk over and shunt the device off!

Incidentally, you can string three 9v DC transistor radio batteries in series to power the unit. The Sonalert can take the resultant 27v DC in stride, although your cheap morning alarm will be all the more alarming for your trouble.

Suggestion: Check with the weatherman the night before. (Otherwise you'll be known at the office as Sunny Day Sam.)

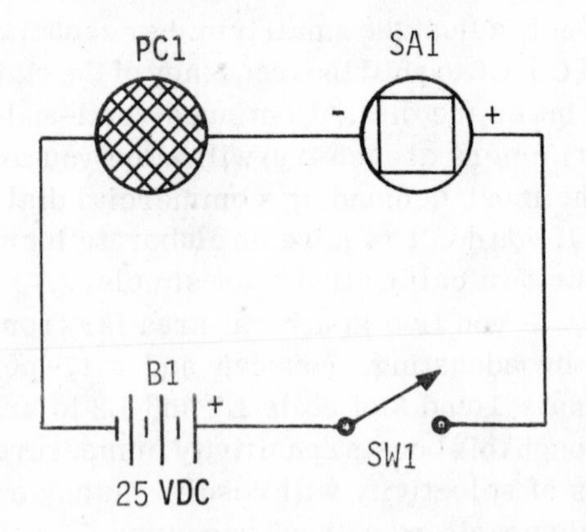

PARTS LIST

B1—22 to 28v DC

PC1—Cadmium sulfide photocell. Lafayette 99 C 6321

SA1—Sonalert. Mallory SC-628

SW1—SPST. Oak Type 175

15

DOUBLE BATTERY SUPPLY FOR THE CAR

Many hobbyists these days are overloading their automobile's electrical system with such accessories as ham radio units, citizens radio tranceivers, cassette recorders, etc., to the detriment of the overall system. One solution, however, is simply to add another battery to the car.

Two high-current, low-voltage silicon diodes can be employed as one-way switches as shown in the accompanying schematic. Be sure to mount the diodes on heat sinks in as cool a place as can be safely accomplished near the batteries. Use heavy interconnecting wire.

There will be a slight voltage drop across the diodes (about one-half volt), so you may wish to readjust the voltage regulator accordingly for a slightly higher output. This may or may not be necessary, however.

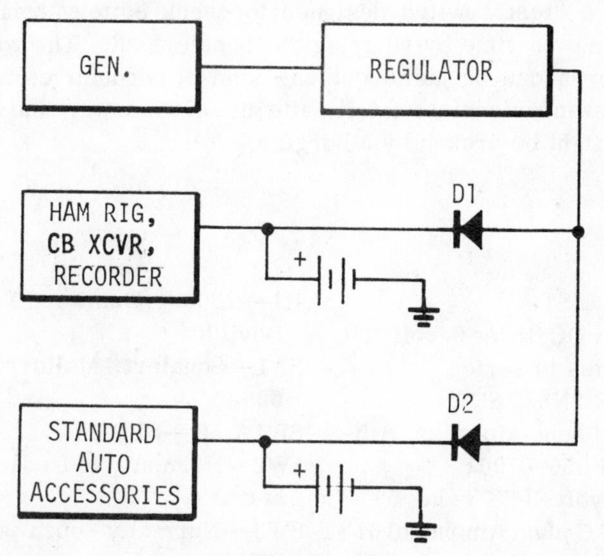

PARTS LIST
D1, 2—1N1396 silicon diodes, mounted in heat sinks.

16

SONALERT INTRUSION ALARM

The new Sonalert devices offer some distinct advantages when employed as an integral part of a home intrusion alarm. For our purposes the prime features are the Sonalert's ability to function well off low-voltage supplies, plus the added advantage of producing a high, penetrating sound that can be heard over some distance.

The handy burglar alarm shown here employs a "fail-safe" double-check system. If the AC current should fail for some reason, the alarm will sound. The only way to stop the alarm is to turn off SW1. In order to start the alarm again (reset) switch SW2 must be momentarily pressed. Switch SW3 is merely a "test" switch designed to check battery condition from time to time by shorting the Sonalert on. The window pane shown can be just about any kind of conductive circuit: door or window contacts, foil patterns on windows, fine wires which might be broken by a burglar, etc.

PARTS LIST

B1—27v DC; three 9-volt TR batteries in series
D1—GE 2N885 SCR
J1, 2—Phone-tip jacks. Amphenol 350-61001
K1—6-volt SPST relay
PL1—AC plug. Amphenol 61-M11
PL2, 3—Phone-tip plugs. H. H. Smith 108
R1—1200. Ohmite "Little Devil"
SA1—Sonalert. Mallory SC-628
SW1A, B—SPDT
SW2—Normally-closed push-button
SW3—Normally - open push-button
T1—6.3v AC filament transformer. Triad F-14X

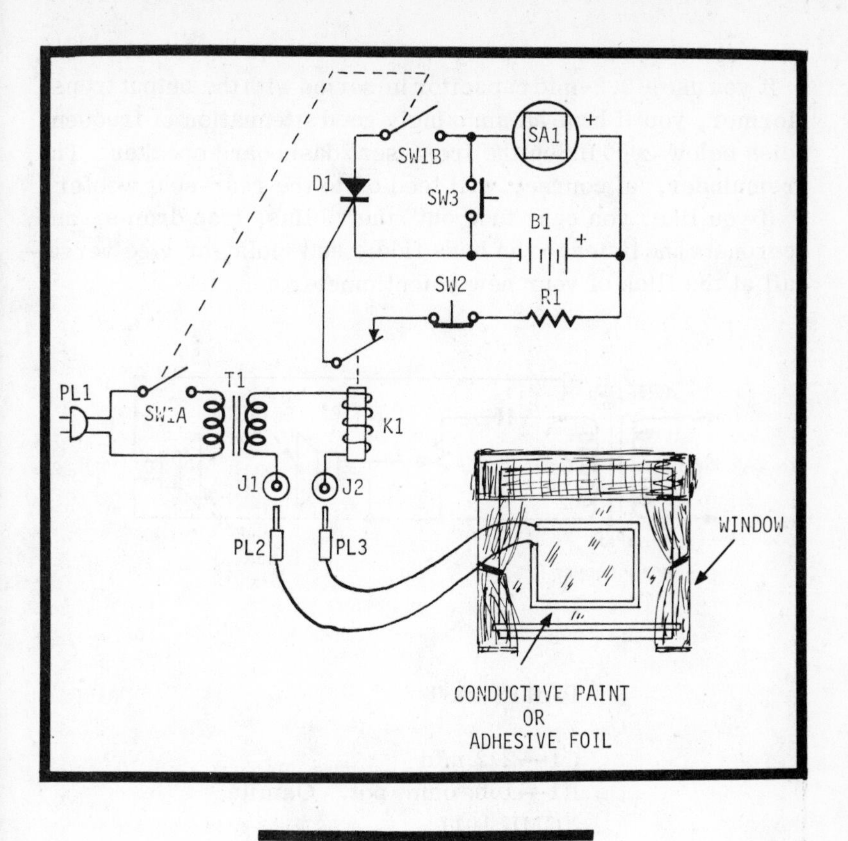

17

HI-FI IN THE CAR

The usual rear-seat speaker arrangement in a car is connected in parallel with the front speaker to provide an on-paper hi-fi effect. Unfortunately, though, this doesn't normally happen because the driver hears more sound from the front speaker than he does from the back, and the base notes invariably are behind him with "tweeter" notes coming from the dashboard. So why not take advantage of this fact?

If you balance both speakers with a 100-ohm potentiometer and equalize the level at the driver's seat, you can obtain a "poor man's hi-fi," providing it's done properly.

If you use a 2.2-mfd capacitor in series with the output transformer, you'll have an amazingly good attenuation of frequencies below 2000 Hz on the front-seat dashboard speaker. The remainder, of course, will feed on to the rear-seat woofer.

If you like, you can "tune out" the violins, trap drums, and coronets and listen to the bass fiddle and viola, or vice versa, all at the flick of your new potentiometer.

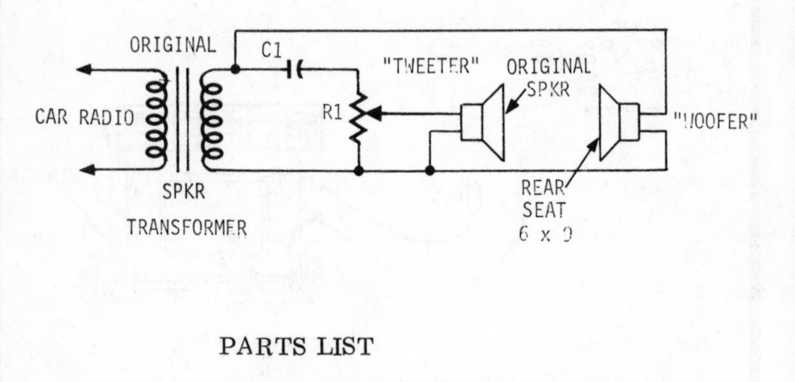

PARTS LIST

C1—2.2 mfd
R1—100 ohm pot. Ohmite CMU 1011

18

A "WHISTLER" LISTENING DEVICE

"Clicks," "chinks," "tweeks," and other weird sounds that frequently resemble a "dawn chorus," abound in the world of radio whistlers. Scientists are still trying to figure out what they are, but at this writing they still don't know much more than they did 20 years ago. Even without elaborate equipment you can hear the mysterious signals if you have a hi-fi phono system with a reluctance pickup head and a separate preamplifier. Combine these with a good-sized loop antenna and the filter circuit shown here, and you'll soon be eavesdropping on some of the most unusual radio signals known to man.

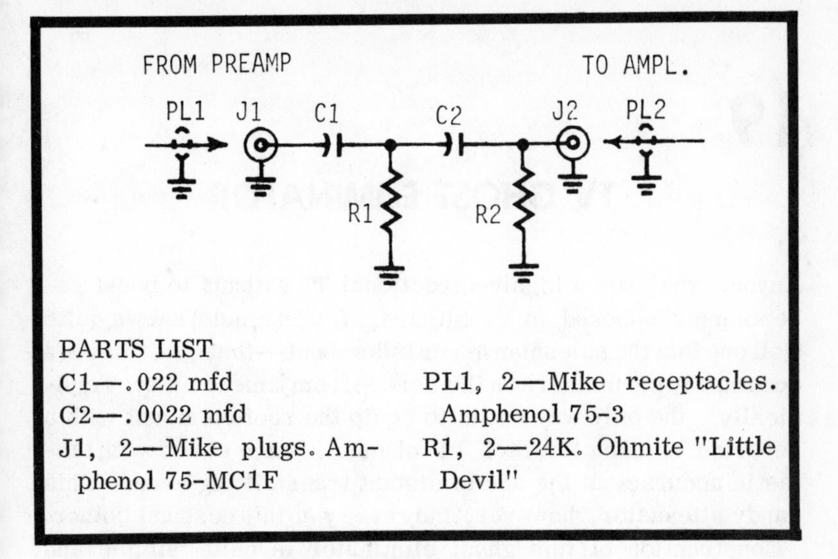

Using any convenient spool of fine, insulated coil wire, wind 20 to 30 turns in a loop around a doorway. This can be easily accomplished by thumbtacking fairly large tacks into the four corners of the door and using this as your "coil form." Next, hook the filter shown in the schematic between the preamplifier and main hi-fi amp. This serves to minimize AC hum pickup.

Now, remove your coil loop from the doorway and suspend it in the same vertical position by means of string taped to the ceiling. Hang it in such a way that the two free end wires of the loop dangle over your record player. Connect the wires to the cartridge input prongs on the record player arm, disconnecting the cartridge temporarily.

Now you're ready for action. Turn on the amplifier (as loud as you want) and see what you can hear. If you still hear some AC hum, simply revolve your suspended loop antenna until the noise is "nulled" out.

The weird "tweeks" and other noises heard faintly as you "tune" the hi-fi's gain control are genuine whistlers. The only thing really known about them is that they seem to come in best during early morning hours and whenever you are within 600 miles or so of a thunderstorm.

19

TV GHOST ELIMINATOR

Anyone who uses a highly-directional TV antenna to boost reception (as opposed to rabbitears, for example) knows quite well one fact the salesman never talks about—that ghost images accompany performers on the screen from time to time. Technically, the only way out is to equip the rooftop mount with a separate antenna for each TV channel, each aimed with fantastic accuracy at the actual distant transmitting tower. This dandy attenuator, however, may save you this cost and bother.

Construction of this ghost eliminator is quite simple and straight-forward. It can be built into any handy Minibox or container and, if you like, switched in and out of the TV feed-line as desired.

To adjust your finished eliminator, set potentiometers R3 and R4 until they are electrically off. Then adjust R1 and R2 until the ghosts are eliminated. Now turn R3 and R4 back on again and adjust them for the clearest picture.

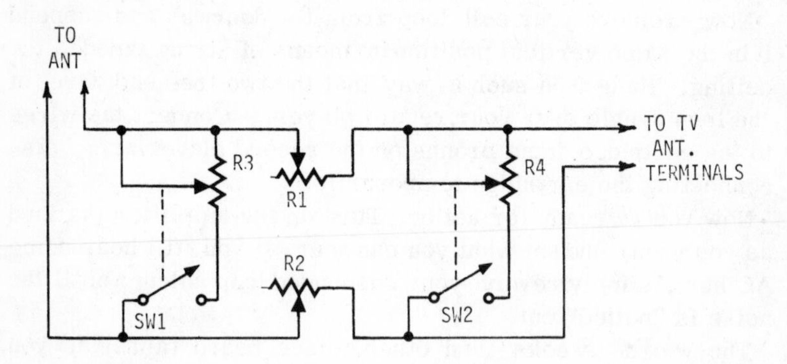

PARTS LIST

R1, 2—10K pots. Ohmite CMU 1031

R3, 4— 2.5K pots. Ohmite CMU 2521

SW1—SPST on R3. Ohmite CS-1

SW2—SPST on R4. Ohmite CS-1

20

ADJUSTABLE NOISE GENERATOR

For aligning and troubleshooting receiver circuits, a useful bit of equipment is a noise generator. Shown in the accompanying diagram is an easy-to-construct version that produces noise from below the AM broadcast band on up through 450 MHz. Output can be adjusted by means of potentiometer R2. Be sure to keep all leads as short as possible.

To align a receiver, feed the output from the generator to the antenna fitting. Now, simply adjust the receiver for minimum noise while retaining maximum sensitivity. You may wish to simply turn the generator on and off while aligning the receiver. If so, you can install an inexpensive single-pole single-throw switch between the battery and R2.

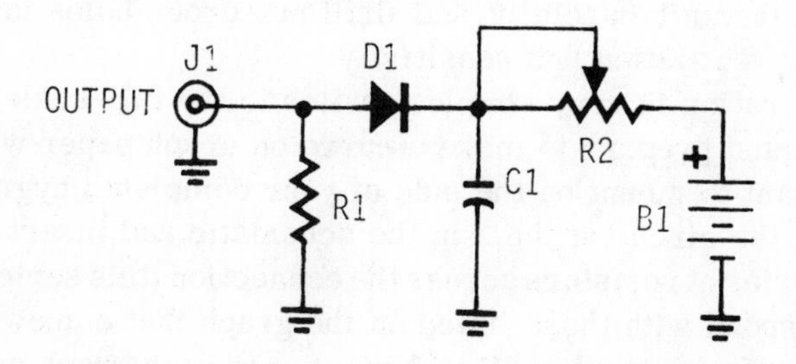

PARTS LIST
B1—3 to 9v DC
C1—.002 mfd
D1—1N21
J1—SO-239 coax fitting. Amphenol 83-1RTY
R1—47. Ohmite "Little Devil"
R2—75K pot. Ohmite CMU 7531

21

PROFESSIONAL HYGROMETER

What's a hygrometer? Essentially, it is a device which makes it possible for you to measure relative humidity (RH) in your home, in the kiddies' room, in the refrigerator, outside the house, or just about anywhere where air moisture is important. Heart of the unit is a plastic sensor unit that is supplied with a resistance vs RH curve that can be used to interpret readings (both Lafayette and Allied Radio supply these with their units). You'll notice that the hygrometer shown here is equipped with a hi-low switch. This permits your reading 30 to 100% on one scale and from about 15 to 32% on the other.

The graph accompanying the sensor shows how the element's resistance changes with varying humidity. For example, at 100,000 ohms across the sensor you're seeing 32% humidity. Although you can remote-line your sensor element, you'll probably want to install it inside the hygrometer. In this case, mount the unit carefully and drill numerous holes to achieve proper ventilation and sensitivity.

Calibration is very simple. Switch to the full-scale reading (High) and prepare to make a curve on graph paper which you will want to mount on the side of your completed hygrometer. Break the circuit at the X in the schematic and insert the first of a series of resistors across the connection (this series should correspond with those listed on the graph that comes with the sensor; for example, 6K to 1 meg). On your graph paper establish RH on the horizontal axis and microamperes on the vertical, so that when you're finished inserting various resistors and taking their readings (checked against the original sensor graph) you have a curve of your own. Now do the same thing after switching over to the Low position, using resistors ranging, say, from 100K to about 1.5 megs.

Take the two graphs and recopy them neatly so that they can be glued to the sides of your hygrometer's case. Spray with Krylon to insure that they don't wear.

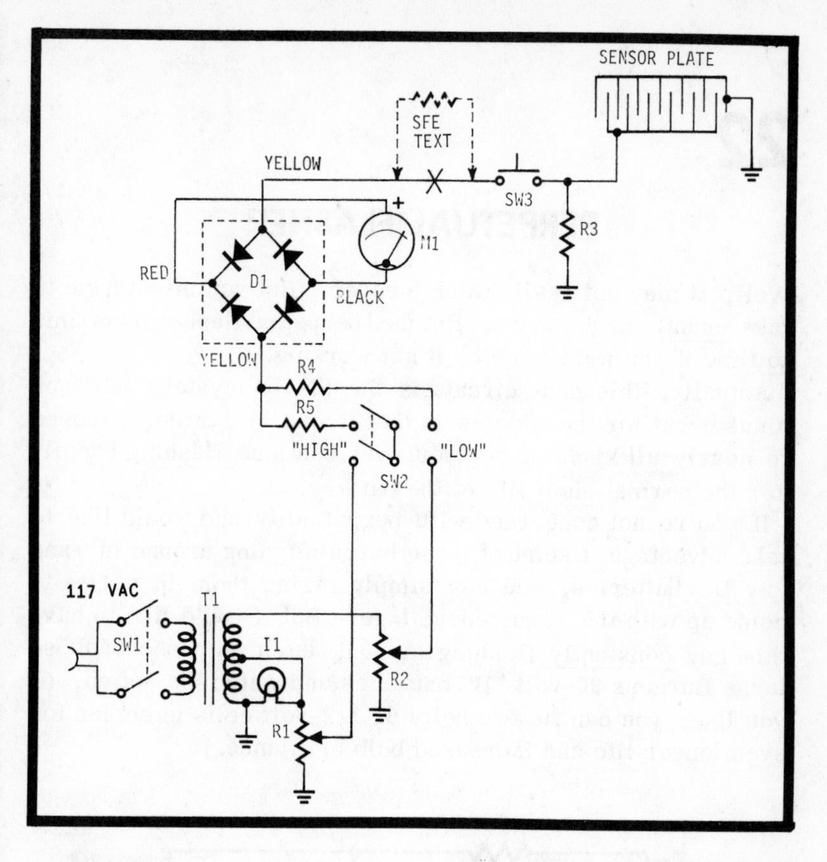

PARTS LIST

D1—Test instrument type bridge rectifier. Conant Labs Series 160B, yellow code

I1—Pilot bulb

M1—0-50 DC microammeter

PL1—AC plug. Amphenol 61-M11

R1, 2—5K pots. Ohmite CMU SO21

R3—110K. Ohmite "Little Devil"

R4—75K. Ohmite "Little Devil"

R5—22K. Ohmite "Little Devil"

SW1—DPST. Oak Type 175

SW2—DPDT. Oak Type 175

SW3—Pushbutton

T1—12.6v AC filament transformer with centertap at 6.3 volts

Now you're ready to go! Always start in the High position, switching over to Low only when the reading indicates an RH below 32%. Out bet is that you'll be right on the nose with the U.S. Weather Bureau every time.

22

PERPETUAL FLASHER

Well, it may not really work forever. But you might have to take annual checks on your finished perpetual flasher from time to time if you want to catch it not working.

Actually, this neat circuit is the ideal "mystery box" or amusement for the kiddies in that it can be jarred, exposed to nearly all kinds of weather, yet continue flashing happily for the normal shelf life of the battery.

If you're not concerned with perpetuality and would like to take advantage of some of those bargains going around in 9 and 22v DC batteries, you can simply string them in series to come up with the required voltage. But if you'd like to have this guy constantly flashing at you, latch onto one of those large Burgess 90-volt "B" batteries and watch the action. (If you like, you can tie two hefty 67 1/2-volt cells in series for even longer life and increased bulb brilliance.)

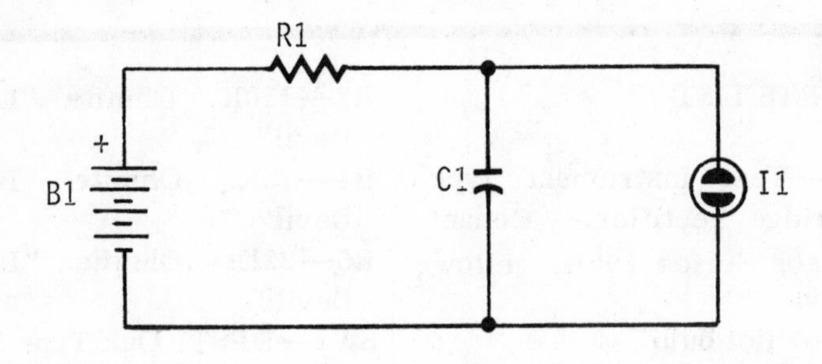

PARTS LIST
B1—90 to 125v DC
C1—.047 mfd
I1—NE-2A
R1—11 meg. Ohmite "Little Devil"

Control for this relaxation oscillator is governed primary by R1, although if you so desire you can substitute a 15-meg or 25-meg potentiometer for the 11-meg shown in the circuit This will permit your adjusting the frequency of flashes. As you decrease the resistance, you increase the flash rate.

23

TONE COMPENSATOR FOR TAPE RECORDERS

Cassette tape recording can be a fascinating hobby in itself, but it's even more exciting when you can actually control the amount of bass or treble that is put onto the tape. The circuit shown here will do this and more. This is the technique employed by professional recording people, although most recorders sold today provide only for your adding bass or treble

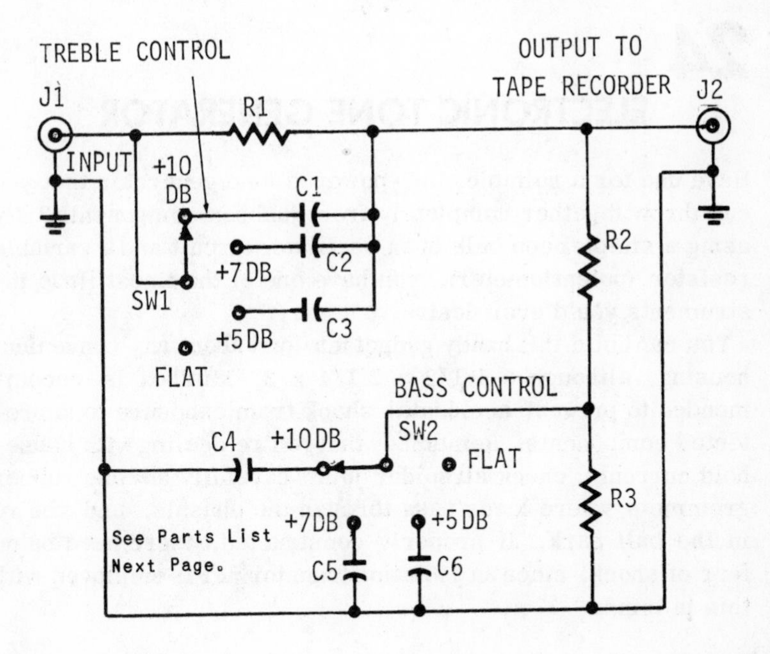

See Parts List
Next Page.

PARTS LIST

C1—260 pfd
C2—510 pfd
C3—.0022 mfd
C4—.022 mfd
C5—.011 mfd
C6—.0047 mfd
J1, 2—Audio receptacles. Amphenol 75-3
R1—510K. Ohmite "Little Devil"
R2—110K. Ohmite "Little Devil"
R3—1.1 meg. Ohmite "Little Devil"
SW1, 2—Four-position rotary switches. Oak Series 20

after the tape has been cut. This is sufficient for most enthusiasts, but for the truly discriminating person it is far from adequate.

If you're putting your records on tape, this unit will allow you to "tune out" the scratches ahead of time. If your records are clean but a mite weak in highs, this gadget will bring them out. If you're crazy about booming bass fiddles, take heed: This gadget will increase bass notes by a full 10 decibels.

24

ELECTRONIC TONE GENERATOR

Have use for a reliable, AC-powered tone generator that you can throw together completely from junkbox components? By using a simple neon bulb in an oscillator circuit and a variable resistor (potentiometer), you have one of the finest little instruments you'd ever desire.

You can build this handy gadget into just about any convenient housing, although a 1 1/2 x 2 1/4 x 3" Minibox is recommended to prevent accidental shock from exposure to unprotected components. Remember that you're dealing with household currents; check all solder joints carefully and use rubber grommets where leads pass through the chassis, and you're in the ball park. If properly constructed, there need be no fear of shock, since an isolation transformer is employed with this in mind.

Once completed, you can use our electronic tone generator to check amplifiers and speakers, to check modulation of your ham or CB rig, or even for code practice!

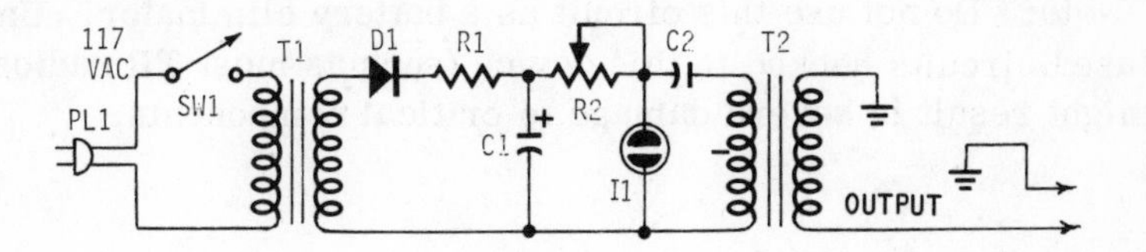

PARTS LIST

C1—40 mfd, 200 WVDC electrolytic
C2—.0047 mfd
D1—25 ma selenium rectifier
I1—NE-2
PL1—AC plug. Amphenol 61-M11
R1—51K. Ohmite "Little Devil"
R2—2.5 meg pot. Ohmite CMU 2552
SW1—SPST. Oak Type 175
T1—117v AC isolation transformer
T2—Transistor radio transformer. Utah 1755

9-VOLT BATTERY CHARGER

In case you're looking for a really good charger/rejuvenator for 9-volt TR batteries and don't want to go around modifying other circuits (such as shown in Project 9), here's a terrific little gem that you can put together in less than 15 minutes if you're at all handy with a soldering gun.

Your junkbox should yield everything you'll need, except perhaps the TR-battery clip, which you can either steal from a junked TR radio or buy from the corner supply store for less than 25¢.

In many respects, this charger is better than most since it has been especially designed for 9-volt batteries. For this reason the charging is accomplished at approximately 17 milliamperes—somewhat conservative for most cells but just the ticket for many nine volters. Do not attempt to charge any other kind of batteries with the device. Depending upon the condition of your battery, it can be brought up to near full capacity in less than an hour or left charging overnight.

Note: Do not use this circuit as a battery eliminator. Unfused circuits hooked to this device (such as most TR radios) might result in severe damage to critical components.

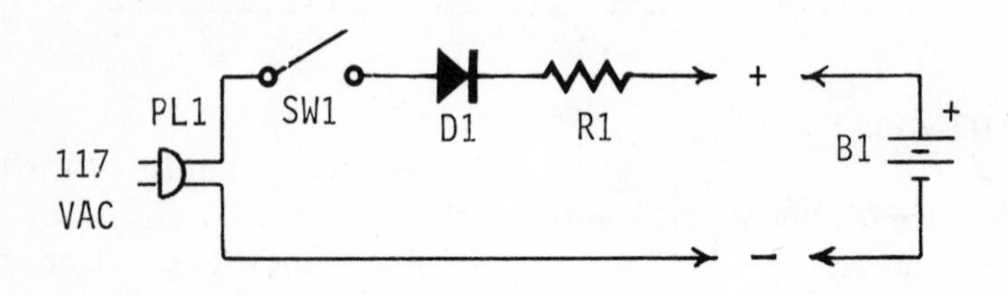

PARTS LIST

B1—9v DC battery under test

D1—200 PIV silicon rectifier

PL1—AC plug. Amphenol 61-M11

R1—4.5K, 8 watts Ohmite "Brown Devil" Type 1544

SW1—SPST. Oak Type 175

26

SUPER FIELD STRENGTH METER

You may think that a field strength meter's a field strength meter. Yet a glance at some commercial surveillance and communications catalogs will confirm that a professionally-made FSM that'll handle the job <u>this</u> one will run into big money!

The trick is in giving the standard FSM a hot front end, achieved in this circuit by critical coils switched in and out and tuned with C1. Why so many bands? Simply because of the common shortcomings most conventional field strength meters display, such as the inability to distinguish between a harmonic and a fundamental frequency, inability to dis-

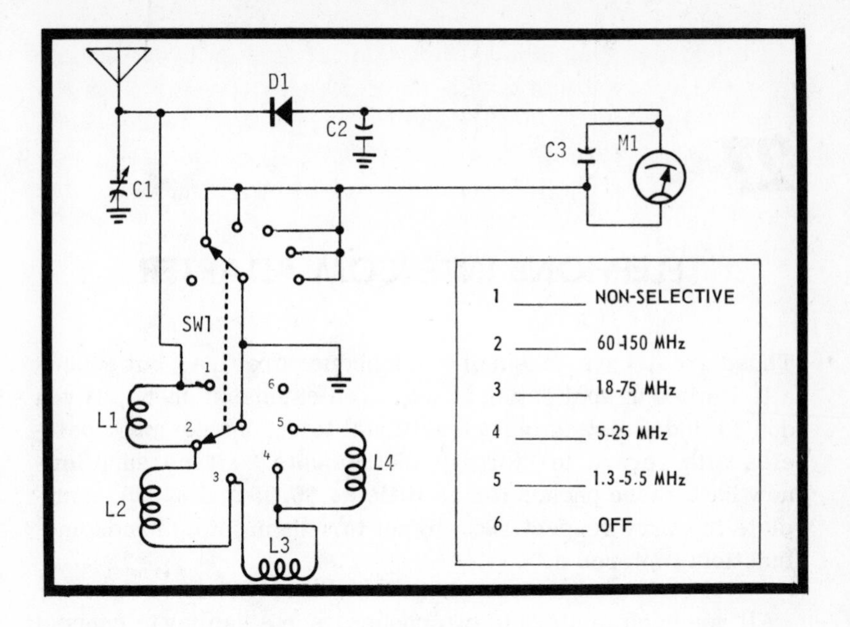

PARTS LIST

C1—360-pfd variable
C2—.0047 mfd
C3—.0022 mfd
D1—1N38B
L1—1/4" diameter single loop of #20 bare hookup wire with leads (which are 1/4" long) spread apart 1/2" at the ends
L2—5 turns of #18 enameled closewound on a 5/16" form
L3—2.4 mh RF choke. J.W. Miller 4606
L4—24 mh RF choke. J.W. Miller 4626
M1—0-50 microammeter
SW1—2 pole, 6-position rotary switch. Oak Type A

tinguish between your signal and the 50,000-watt AM broadcast station 10 miles away, etc.

This circuit, actually a well-designed absorption wavemeter, will cross this barrier in accordance with the following frequency ranges: 1.3-5.5 MHz, 5-25 MHz, 18-75 MHz, and 60-160 MHz. Should you desire wide-open FSM operation, simply flick the switch to position 1 and you're in business.

Coils should be wound exactly as specified in the parts list, although a few liberties can be taken if you have a grid dipper handy. Mount the coils right on the switch.

27

TELEPHONE INTERCOM ADAPTER

Those great buys in surplus telephones are fine, but generally limited in application to use as an extension phone, if you don't mind violating most traditional telephone company policies with regard to "foreign instruments." Many suppliers now have these phones for as little as $6.95 and $7.95, complete in every respect. So why not turn them into a handsome, functional intercom.

All you need to start is two phones, some cabling to connect them with, and the intercom adapter shown in the accompanying schematic diagram. The adapter connects to the yellow, green, and red wires and furnishes all the power you'll need to realize full operation of the phones. Many people prefer to omit SW1 altogether, since current drain when the phones are not used is nearly nil.

Using the components suggested here, you can power up to three telephones, all connected together on a party-line circuit. You can assign each phone with a special code. (For example: 3 rings for phone three, 1 ring for phone one, etc.) To get your party, simply dial the number corresponding to the number of times you want the phones to ring.

PARTS LIST

C1—1000 mfd, 25 WVDC electrolytic

D1, 2—30 PIV, 500 ma silicon rectifier

K1—6v DC SPDT relay. Potter & Brumfield KA5DY

PL1—AC plug. Amphenol 61-M11

R1—43. Ohmite "Little Devil"

SW1—SPST. Oak Type 175

T1—125 and 6.3v AC transformer

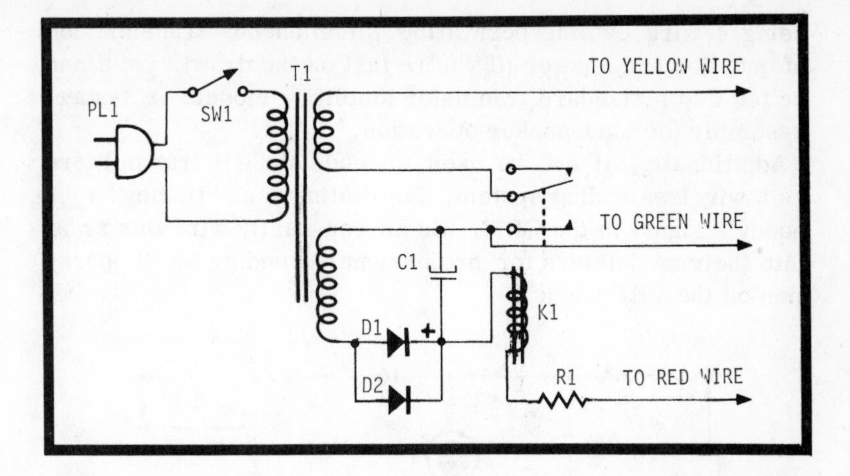

Remind all phone users that they must not answer the instrument until the phone has stopped ringing. The reason for this is that you must allow time for relay K1 to remove the ring current after dialing stops!

28

NEON BULB OSCILLATOR

While several applications of neon bulb relaxation circuits are presented in this book, here is the raw configuration itself—just in case you'd like to apply it to a gadget of your own design. Essentially, this arrangement starts the neon tube flashing very rapidly, although you might not guess that it is intermittent by simply watching the bulb. The flashing-rate is quite high, causing an audible oscillation of tone when monitored by a pair of headphones.

If you take it strictly as presented here, you'll have a portable oscillator adaptable to any number of purposes. If you like, you can make a field code telephone system out of it, using 4-wire cabling permitting simultaneous transmission of question and answer (if you're fast on the draw). Or it can be fed into a standard transistor amplifier module or 3-stage assembly for loudspeaker operation.

Additionally, it can be used to modulate CB transmitters as a wireless coding system, facilitating your "finding" your buddy's signal in the QRM. Hams can easily wire this as is into their modulators for professional-sounding MCW operation on the VHF bands.

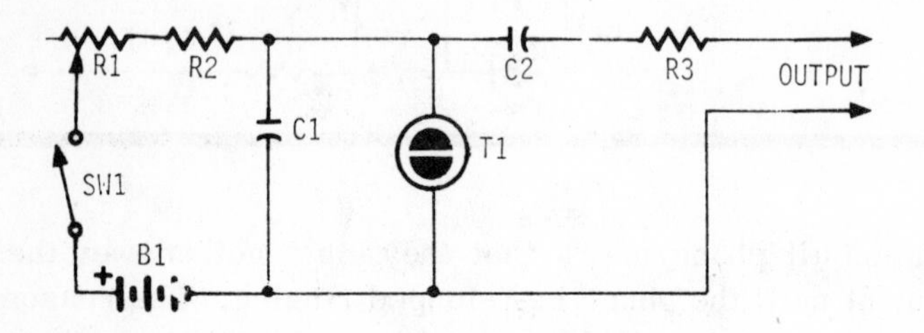

PARTS LIST

B1—90v DC
C1—.0047 mfd
C2—.15 mfd
I1—NE-2
R1—1 meg pot. Ohmite CMU 1052
R2—160K. Ohmite "Little Devil"
R3—510K. Ohmite "Little Devil"
SW1—SPST on R1. Ohmite CS-1

29

ADD-ON NOISE LIMITER

Here's a jim dandy of a circuit, one which many manufacturers are now considering incorporating into some of their top-of-the-line communications receivers and CB gear. Unlike most conventional noise-limiting circuits, this one takes the audio going to your headset, removes the noise, and sends only the pure signal on its way.

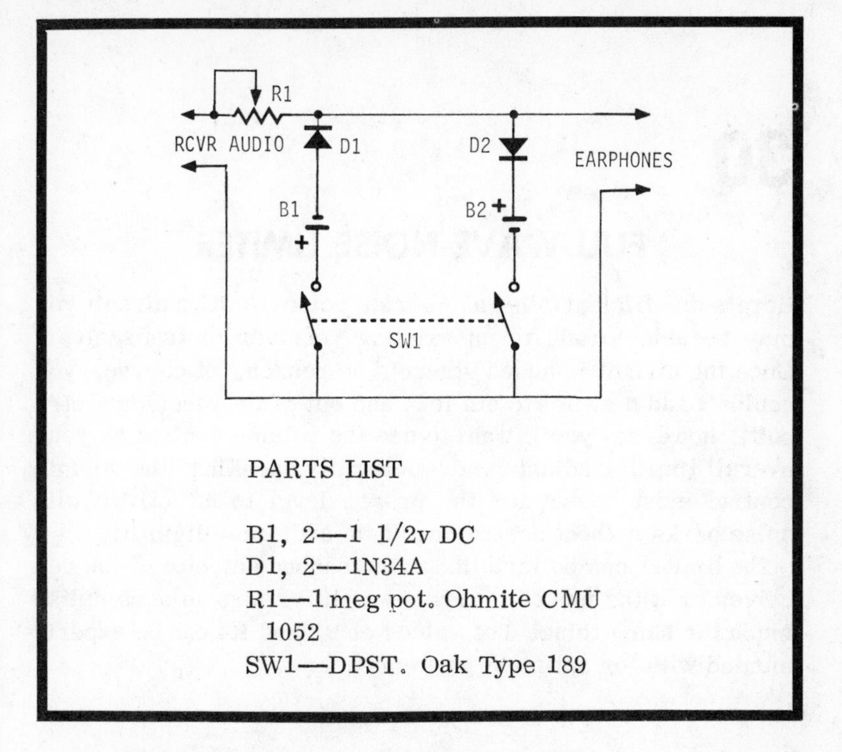

By effectively clipping out loud noise pulses, you hear only the steady signal you're actually after and not flourescent buzzing, ignition noise, and the rest of man-made interference that often seems unsurmountable to the eager ham, CBer, or SWL.

You can build your noise limiter in a small Minibox equipped with jacks to accommodate your headset plug and a line to the receiver. You'll find you can vary the amount of clipping by simply adjusting R1 to a point where best noise reduction is realized.

Although primarily designed for headset operation, you can use this on a speaker if you keep the volume down. Better yet, install an audio amplifier (a one or two-transistor cheapy will do) inside the speaker enclosure and feed the output of your noise limiter to it. In this way you won't overdrive the crystal diodes, yet you'll still be able to enjoy room-filling volume.

30

FULL-WAVE NOISE LIMITER

Here's an efficient full-wave series noise-limiting circuit you may be able to add to an existing receiver or transceiver. Once the circuit is added (there's no reason, of course, you couldn't add a switch to cut it in and out of the receiver's circuit), however, you'll want to use the volume control as your overall limiting adjustment. Generally speaking, the volume control must be set for the proper level to effectively clip noise peaks without destroying basic audio intelligibility.

The limiter can be installed at the second detector of the receiver or at the input to an audio amplifier stage to accomplish much the same thing. The values of R2 and R4 can be experimented with for optimum performance.

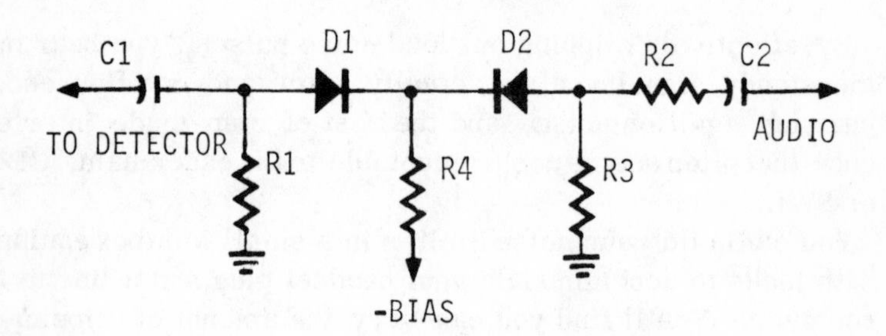

PARTS LIST

C1, 2—.047 mfd
D1, 2—1N56
R1, 3, 4—1 meg. Ohmite "Little Devil"
R2—100K. Ohmite "Little Devil"

31

CB-TO-AUTO ANTENNA ADAPTER

One of the handiest little gadgets we've ever come across is this simple circuit, borrowed from a commercial antenna matcher we opened after plunking down $6.95 for it. So what was inside? Perhaps $1.75 worth of parts—if everything was purchased brand new.

What this gem does is permit you to use a standard AM car radio antenna as a CB transmit/receive antenna, thereby saving you the expense of a second 27-MHz radiator and the aggravation of drilling holes in the family buggy. No, the car radio antenna isn't as efficient as a full-size CB whip, but it will do an admirable job if properly matched to the rig.

We simply built ours into a clear area in the rear of the glove compartment, although you may want to enclose it in a small plastic box to be tucked away behind the car radio. Use standard auto antenna plugs and jacks and you'll save yourself a lot of headaches, although you'll probably need one PL-259 to couple to the output connector on your CB rig.

Once installed, hook an SWR bridge into the feedline and throw the transmit switch on the CB rig. With an eye on the meter, adjust the height of the telescoping car antenna for minimum SWR and leave it there. Next, tune L1 for the same thing (minimum reflected power). Now retune the final loading control a mite and repeat the above process until the absolute minimum has been reached. If you still find SWR a bit high (say, above 2.5:1), shorten or lengthen the coax between the adapter and the CB rig and repeat the tuneup procedure. Finally, mark this spot on the telescoping antenna with a thin ring of electrical tape or other means of marking so that you'll always know the proper height to set the antenna.

Lastly, readjust the AM car radio front end by tuning the trimmer capacitor for best overall performance with the adapter. No, you don't have to open the radio. This trimmer is generally accessible through a small opening on the side of the radio near the antenna jack.

If you notice an increase in ignition noise on the car radio or CB set, shield the adapter by enclosing it (if you have room) inside the CB unit. Better still, build it into a tiny metal Minibox in the first place and run a ground strap to a convenient grounding point under the dashboard. This should completely eliminate the noise.

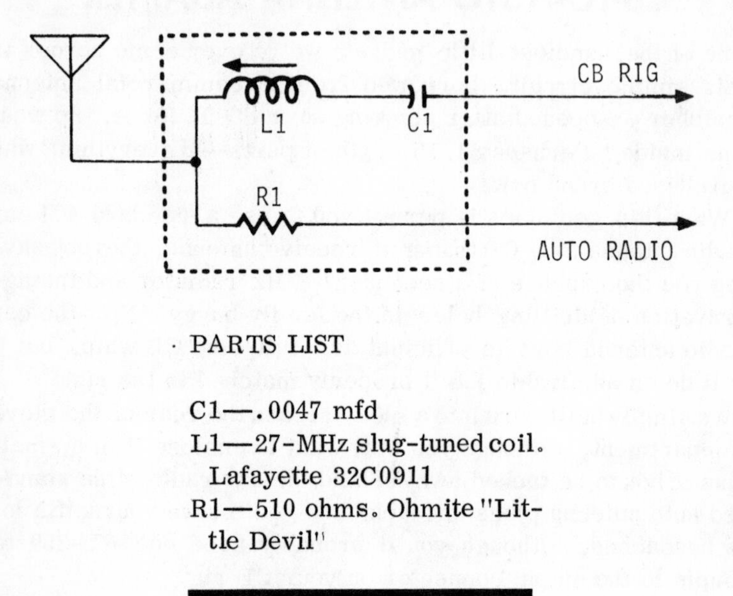

PARTS LIST

C1—.0047 mfd
L1—27-MHz slug-tuned coil. Lafayette 32C0911
R1—510 ohms. Ohmite "Little Devil"

32

AUDIO WATTMETER

Here's a gadget you won't find in the stores, yet it will accurately measure your hi-fi's output in watts automatically by a flash of a bulb!

Nothing is critical about construction, although it should be mentioned that potentiometers R1, R3, R5, and R7 can be mounted inside the housing, with only the neon lamps and a pair of wires with alligator clips on the ends protruding.

To calibrate your wattmeter, adjust the output voltage of the calibrator supply to 11 volts (as measured on voltmeter) and adjust R7 so that I4 just lights as this voltage is applied

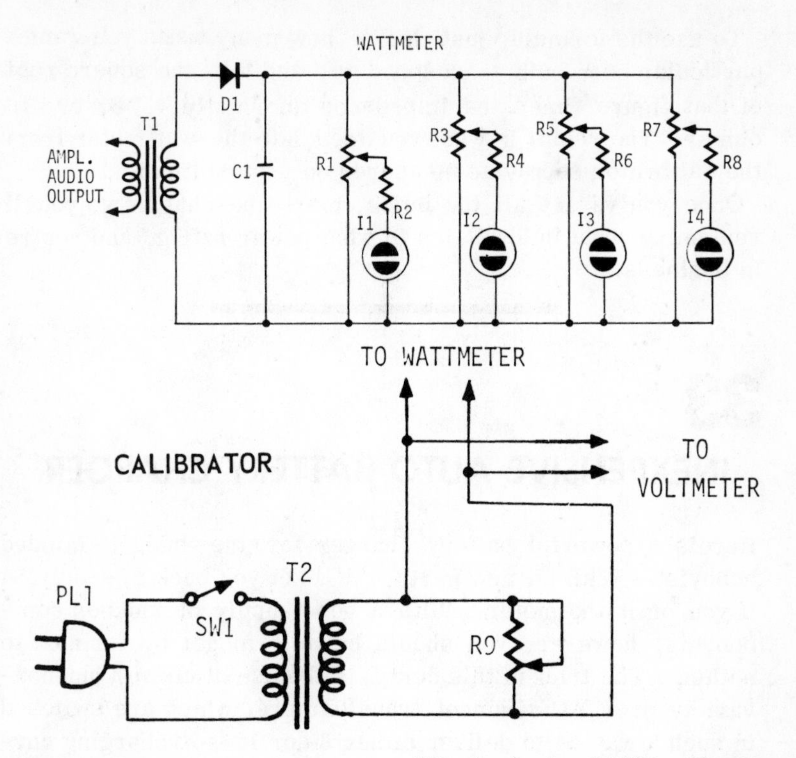

PARTS LIST

C1—.22 mfd

D1—100 ma, 400 PIV silicon rectifier. Lafayette SP-196

I1, 2, 3, 4—NE-2

PL1—AC plug. Amphenol 61-M11

R1, 3, 5, 7—1 meg pots. Ohmite CMU 1052

R2, 4, 6, 8—110K. Ohmite "Little Devil"

R9—150 pot. Ohmite CMU 1511

SW1—SPST. Oak Type 175

T1—Universal Output transformer. Lafayette TR-12

T2—12.6V AC filament transformer

to the wattmeter input. Now I4 will light whenever 15 watts of audio is received. Bear in mind, of course, that all bulbs up to the one that indicates the input power will light. Conversely, you can set I1 for 3-watt firing by feeding 5 volts into the wattmeter and adjusting R1 so that the bulb just lights.

For your own settings, use the formula E equals W x Z, where E is voltage you control by adjusting R9, W is power in watts, and Z is the amplifier's output impedance. (If your hi-fi feeds 8-ohm speakers, Z is 8.)

To use the formula, just decide how many watts you want a particular neon bulb to respond to, and take the square root of that figure times the impedance (normally 4, 8, or 16 ohms). The result is what you feed into the wattmeter from the calibrator supply as measured on your voltmeter.

Once you've set all the bulbs, mark the chassis so you'll remember what bulb relates to what power rating, and you're in business.

33

INEXPENSIVE AUTO BATTERY CHARGER

Here's a powerful battery charger for the budget-minded hobbyist. With all new parts, it'll set you back $7—only $4 if you omit the meter. With a good supply of junkbox components, however, you should be able to get by for next to nothing. The trick in this design is the use of cheap, junkbox-variety 6.3v AC filament transformers, which are switched in such a way as to deliver either 6- or 12-volt charging currents for the widest possible variety of applications. Though not designed to compete with $60 booster/chargers, this charger will deliver one heck of a lot more power than some $15-$25 trickle chargers, although if you attempt to overdraw from the supply (say, if you want to rejuvenate a nearly-completely-exhausted 12-volt cell), you will have to temporarily short the circuitbreaker until the battery begins to come back to life and causes less drain.

Allow plenty of ventilation holes for the diodes, although you'll want to insure that no moisture gets into your charger. For this reason you may want to form bent-lip vents instead of holes which will allow rain, snow, etc., to merely run off.

When you notice that M1's needle has returned to a near-normal (no-load) position, your battery is probably fully charged. No damage will result if you leave the charger on overnight. Always check the battery cells' fluid levels prior to charging and after you've removed the charger. A fully rejuvenated battery is indicated by a gentle bubbling in each

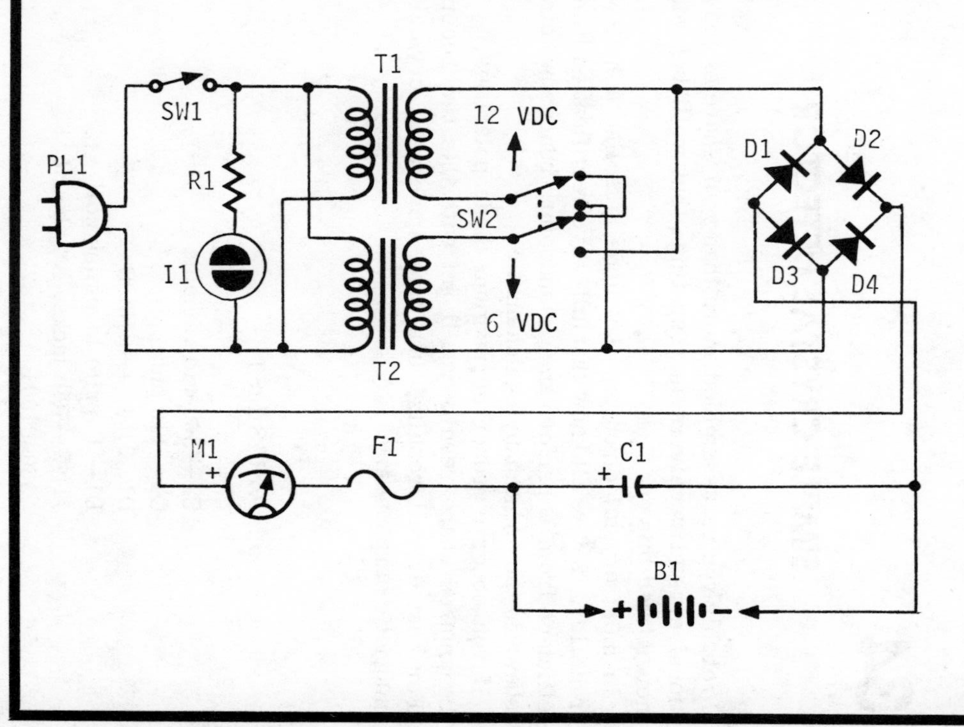

PARTS LIST

B1—Auto battery under charge
C1—80 mfd, 20 WVDC electrolytic
D1, 2, 3, 4—55 PIV, 2-amp silicon rectifiers
F1—Circuitbreaker. Mel-Rain 2-A
I1—NE-2
M1—0-100 DC milliammeter
PL1—AC plug. Amphenol 61-M11
R1—110K. Ohmite "Little Devil"
SW1—SPST. Oak Type 175
SW2—DPDT. Oak Type 200
T1, 2—6.3v AC filament transformer. Triad F-14X

cell during charge. (Always remove the battery caps during charge, replacing them when M1 indicates the job is finished.)

Incidentally, in case you forget to check, the circuitbreaker will disconnect the charger when the job's done. Should it break open early in the game, it may be a sign that the battery's beyond hope. In any case you can short it as indicated earlier, but if this repeats itself after a couple of tries (circuitbreaker still won't hold), the battery has had it.

34

SIMPLE CRYSTAL DETECTOR

Probably you've assembled more than your share of conventional crystal receivers in your time, but we'll bet you've never tried this one!

In most crystal circuits, the diode acts like a half-wave rectifier. Yet we all know that full-wave rectification is more efficient from a performance standpoint, hence the circuit shown in the accompanying schematic.

If you observe proper germanium diode polarities, you'll be amazed at the results you'll get with this tiny receiver. For best overall results, use high-impedance crystal or magnetic earphones.

PARTS LIST

C1—360-pfd variable
C2—110 pfd
D1, 2, 3, 4—1N38B
L1—Ferrite loopstick coil
J1, 2—Tip jacks. Amphenol 350-61001

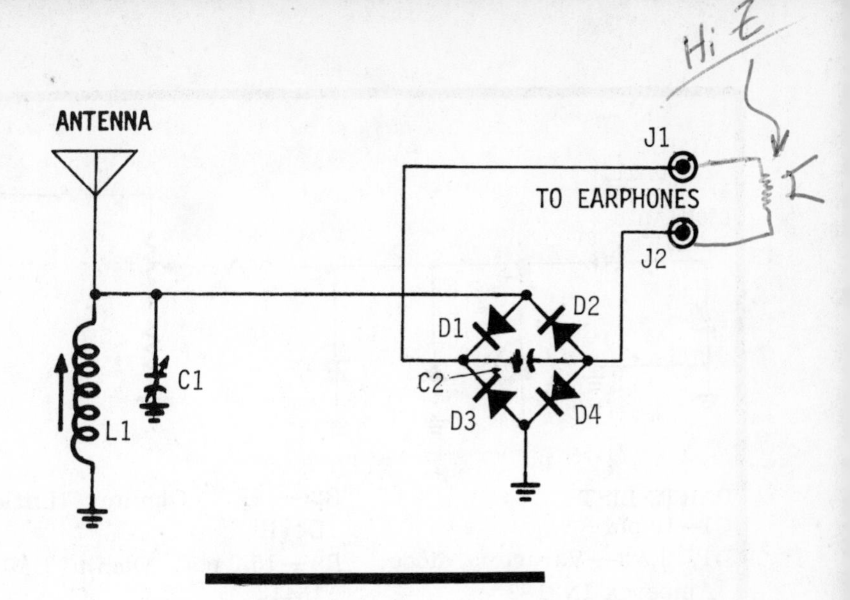

35

SUPER BAND SPREADER

Here's a project that's really a modification, as opposed to a separate gadget you build into a Minibox. Using very inexpensive Amperex 1N3182 varactor diodes, this circuit can make the most battered, inexpensive shortwave receiver sound like a $400 model just taken off the shelf.

Frequently it is next to impossible to find a particular DX station listed in one of the hobby magazines because of other stations that appear to be cluttered around the same spot on the dial. The key is in spreading out the stations so that you can separate them more clearly as you're tuning the dial. Yet this conversion is by no means limited to shortwave receivers; it will do wonders for CB transceivers as well.

Once installed, you'll have a new control to fool with whenever the going gets rough—potentiometer R3. This will serve as a "fine-tuning" adjustment and will make all the difference in the world when it comes to hearing those weak ones.

Incidentally, don't worry about B-plus voltages. The varactors can handle currents found in all receivers.

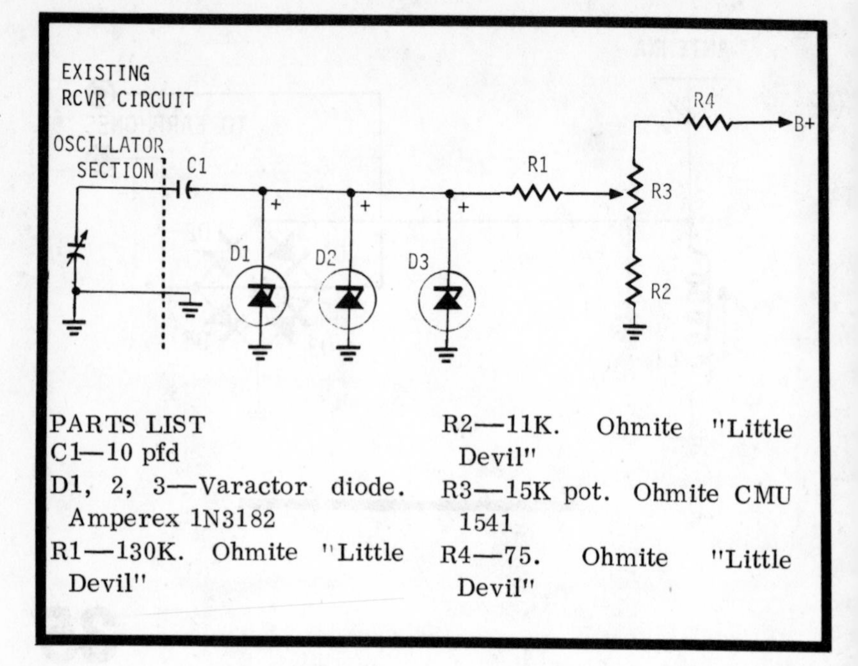

PARTS LIST
C1—10 pfd
D1, 2, 3—Varactor diode. Amperex 1N3182
R1—130K. Ohmite "Little Devil"
R2—11K. Ohmite "Little Devil"
R3—15K pot. Ohmite CMU 1541
R4—75. Ohmite "Little Devil"

TUNNEL DIODE OSCILLATOR

Here's a simple project to construct, yet one that should provide hours of fun for the experimenter who would like to see what a tunnel diode can do.

Shown in the accompanying schematic is the circuit for an easy-to-build broadcast-band oscillator. Coil-capacitor combination C1/L1 resonates over the entire AM band and the frequency can be adjusted by simply tuning the 365-pfd capacitor to the desired spot. To demonstrate its operation, simply tune in a fairly weak AM station on a nearby receiver. After turning SW1 on, hold the tunnel diode oscillator near the broadcast antenna and tune C1 until a whistle is heard. At this point the oscillator is tuned to the frequency of the received station.

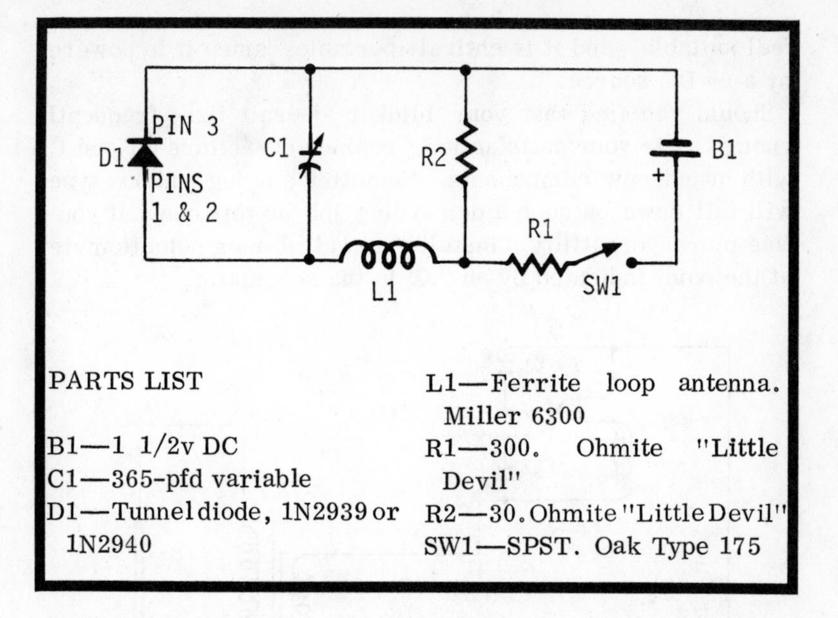

PARTS LIST

B1—1 1/2v DC
C1—365-pfd variable
D1—Tunnel diode, 1N2939 or 1N2940
L1—Ferrite loop antenna. Miller 6300
R1—300. Ohmite "Little Devil"
R2—30. Ohmite "Little Devil"
SW1—SPST. Oak Type 175

Bear in mind, of course, that the unmodulated signal from the oscillator will not be audible unless the receiver is tuned to a weak AM station. The heterodyning action of the two signals beating together produces the audio tone.

37

SIGNAL BLINKER

Ever wish you had a truly brilliant emergency flasher for use on the highway when the car is disabled? Or perhaps a signal light to mount near the driveway or sidewalk after you've finished laying wet cement? Well, this blinker is guaranteed to top them all.

Unlike conventional flashers, this thing is designed for maximum efficiency out-of-doors. It can be built in any form you feel suitable, and it is entirely portable, since it is powered by a 6v DC source.

Should you find that your blinker doesn't fire frequently enough or to your satisfaction, replace capacitors C1 and C2 with brand-new components. Sometimes aging junkbox types will fall down on such a demanding job as this one. If you'd like more versatility, install a 2-watt 1-meg potentiometer at the point indicated by an "X" in the schematic.

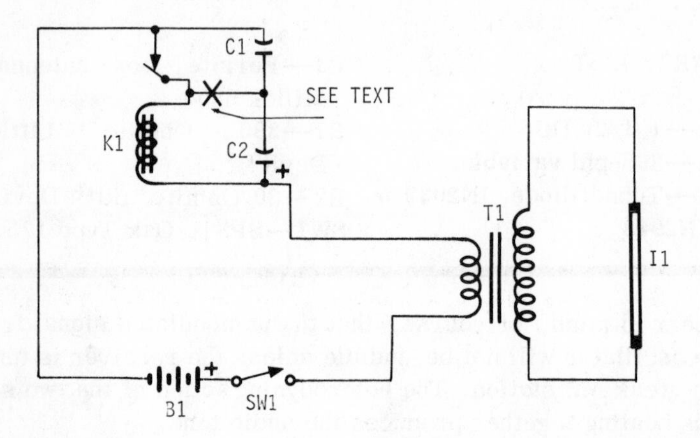

PARTS LIST

B1—6v DC
C1—.3 mfd
C2—2500 mfd, 25 WVDC electrolytic
I1—Red flourescent tube
K1—SPST 100-ohm relay, normally-closed contacts
SW1—SPST. Oak Type 175
T1—Audio output transformer

38

LOUDSPEAKER-TO-MIKE CONVERTER

Here's an extremely useful gadget to have around when you get caught in a pinch without a good microphone. It permits your converting a permanent magnet speaker into a microphone with amazing efficiency. Incidentally, this is the same

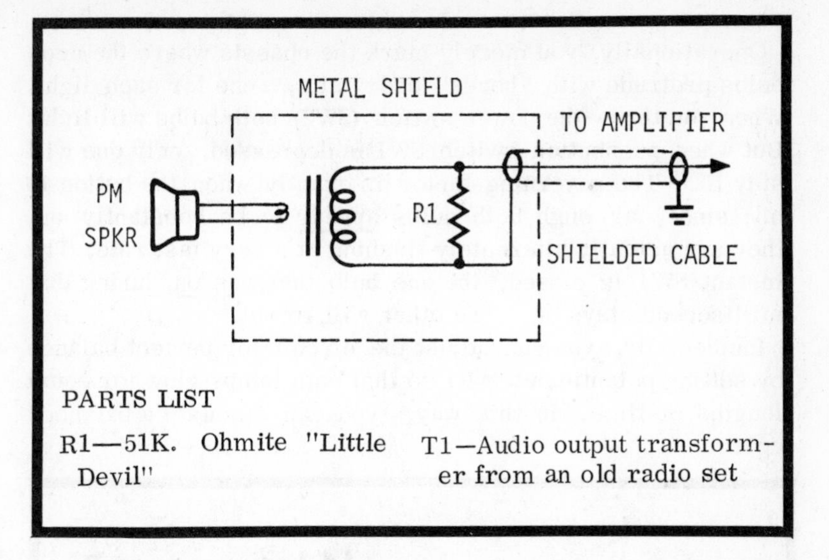

PARTS LIST

R1—51K. Ohmite "Little Devil"

T1—Audio output transformer from an old radio set

circuit used by many walkie-talkie manufacturers in their top-of-the-line 2- to 5-watt models.

The transformer can be the kind used as an output transformer in small radio receivers. Be sure to shield the works as shown in the diagram, however, or an annoying hum will be induced into the modulator or amplifier you are using.

Although the circuit will work without R1, you will find that better frequency response is possible with it in the circuit. If you like, you can substitute a 100K potentiometer and adjust it until tonal quality and response is exactly to your liking.

39

ELECTRONIC "COIN TOSS"

Even though it's common procedure to call out either heads or tails and then flick the coin to decide a debate, arrive at a decision, etc., the experts have proven that heads will come up more frequently than tails if tested for 100 tries. So what? Well, this gadget is electronically programmed <u>never</u> to make an error; to be fair and square all the time.

Operationally, you merely mark the chassis where the neon bulbs protrude with "heads" and "tails," one for each light. When you throw the power switch (SW2) both bulbs will light. But when pushbutton switch SW1 is depressed, only one will stay lit. The governing factor is exactly when the button is hit; since, al ough both bulbs appear to be constantly on, they are actually alternately flashing at a very fast rate. The instant SW1 is closed, the one bulb that was on during that millisecond stays on. The other will go out.

Incidentally, you can adjust the circuit for perfect balance by setting potentiometer R1 so that both lamps glow for equal lengths of time. In this way, you can simulate a balanced coin.

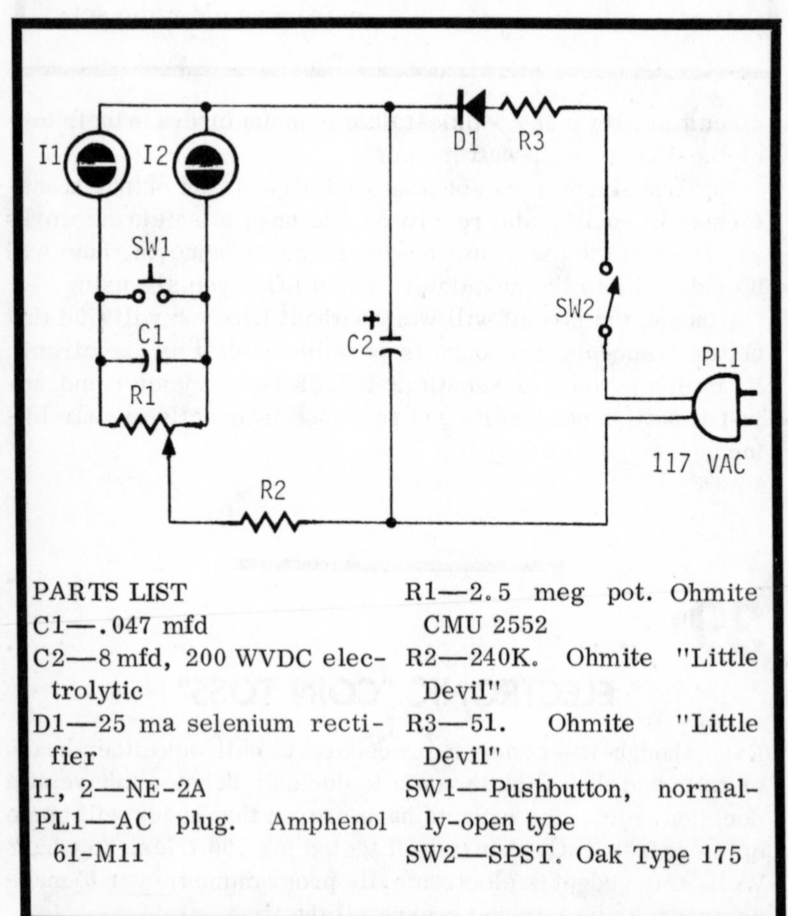

PARTS LIST
C1—.047 mfd
C2—8 mfd, 200 WVDC electrolytic
D1—25 ma selenium rectifier
I1, 2—NE-2A
PL1—AC plug. Amphenol 61-M11
R1—2.5 meg pot. Ohmite CMU 2552
R2—240K. Ohmite "Little Devil"
R3—51. Ohmite "Little Devil"
SW1—Pushbutton, normally-open type
SW2—SPST. Oak Type 175

Construction is simple and straight-forward. Just observe proper polarities, and make certain your NE-2A bulbs are good.

40

INEXPENSIVE VOLTAGE TRIPLER

Here's a simple way of achieving high voltage without the trouble and cost of buying hefty power transformers and rectifiers. The circuit shown here will triple the AC voltage of anything appearing at the secondary of T1 (providing you don't exceed 35 volts) and provide you with rectified DC ready for any application you have in mind. The best part, however,

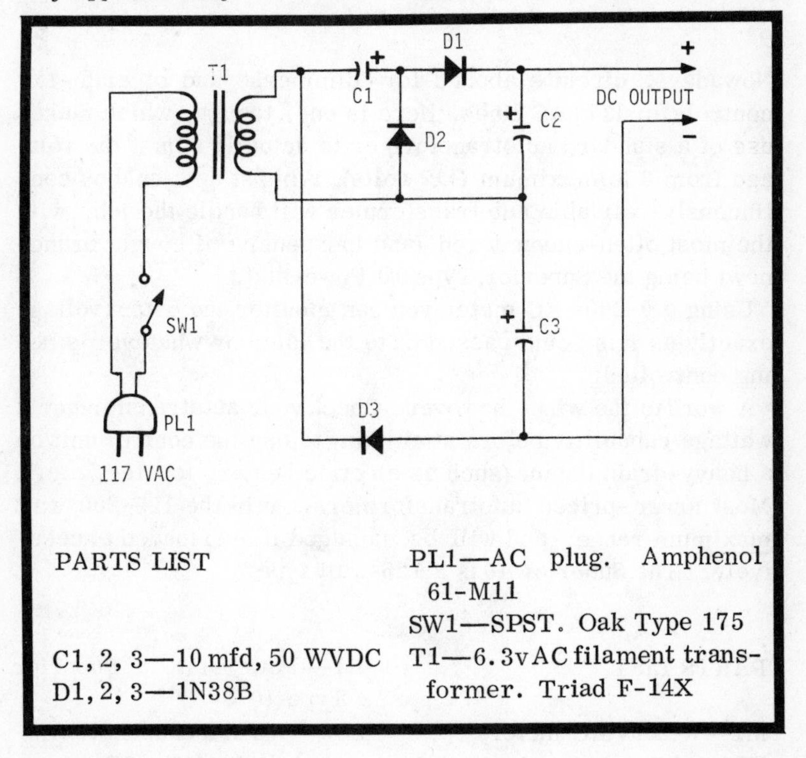

PARTS LIST

C1, 2, 3—10 mfd, 50 WVDC
D1, 2, 3—1N38B
PL1—AC plug. Amphenol 61-M11
SW1—SPST. Oak Type 175
T1—6.3v AC filament transformer. Triad F-14X

is that it uses junkbox-variety 1N38B germanium diodes and a standard 6.3v AC filament transformer.

In the circuit shown you'll wind up with approximately 18v DC output. Other transformers could be substituted for T1, with, of course, proportionately higher or lower DC outputs.

Make sure you observe proper diode, and capacitor polarities and you'll have no trouble whatever with this gadget.

41

LINE VOLTAGE CONTROL UNIT

Nowadays, circuits abound for "dimmers" and other hi-low controls for 117v AC lines. Here is one, though, which makes use of a standard autotransformer to actually adjust the voltage from 0 to maximum (117 volts). Almost any junkbox continuously-variable autotransformer will handle the job, with the most often-encountered (and inexpensive if bought brand-new) being the Superior Type 10 Powerstat.

Using a 0-200v AC meter you can monitor the output voltage exactly as it is being passed on to the lamp or whatever is being controlled.

A word to the wise, however. Check your autotransformer's wattage capability before attempting to use the control unit on a heavy-drain device (such as electric heater, toaster, etc.). Most lower-priced autotransformers are in the 125-200 watt maximum range, and will be damaged if overloaded excessively. The Superior 10 is a 125-watt type.

PARTS LIST

M1—0-200v AC meter
PL1—AC plug. Amphenol 61-M11
R1—Powerstat. Superior Type 10
SO1—AC output socket. Amphenol 61-M1P-61F
SW1—DPST. Oak Type 200

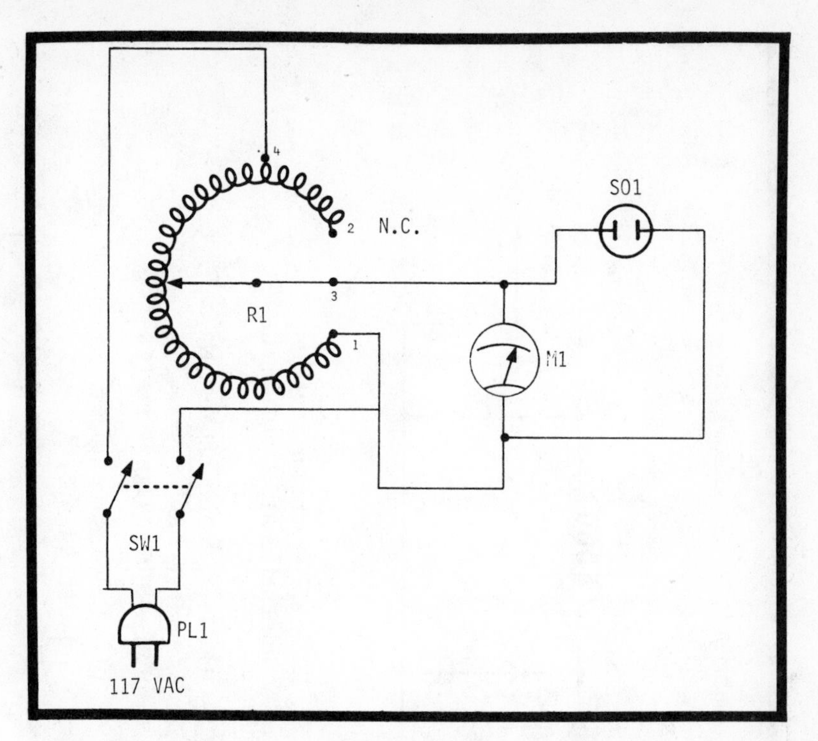

Use heavy wire throughout, making certain that in no instance does the wiring (which should be insulated) come in contact with the metal housing.

If you have in mind the dimming of several low-wattage room lamps, you can add more AC sockets to the control unit, which can be unobtrusively hidden behind the couch.

42

1000-VOLT BUZZER POWER SUPPLY

Using standard junkbox components plus a conventional 6-volt high-frequency buzzer, this project provides an almost unbelievable 1000 volts output at 60 to 70 microamperes DC. It's ideal for science lab experiments, geiger counter power supplies, fence chargers, shock rods, etc., and you can probably put the whole thing together in less than two hours.

The power supply is driven by four Size D flashlight cells wired in series. The unit will provide a constant 1000-volt

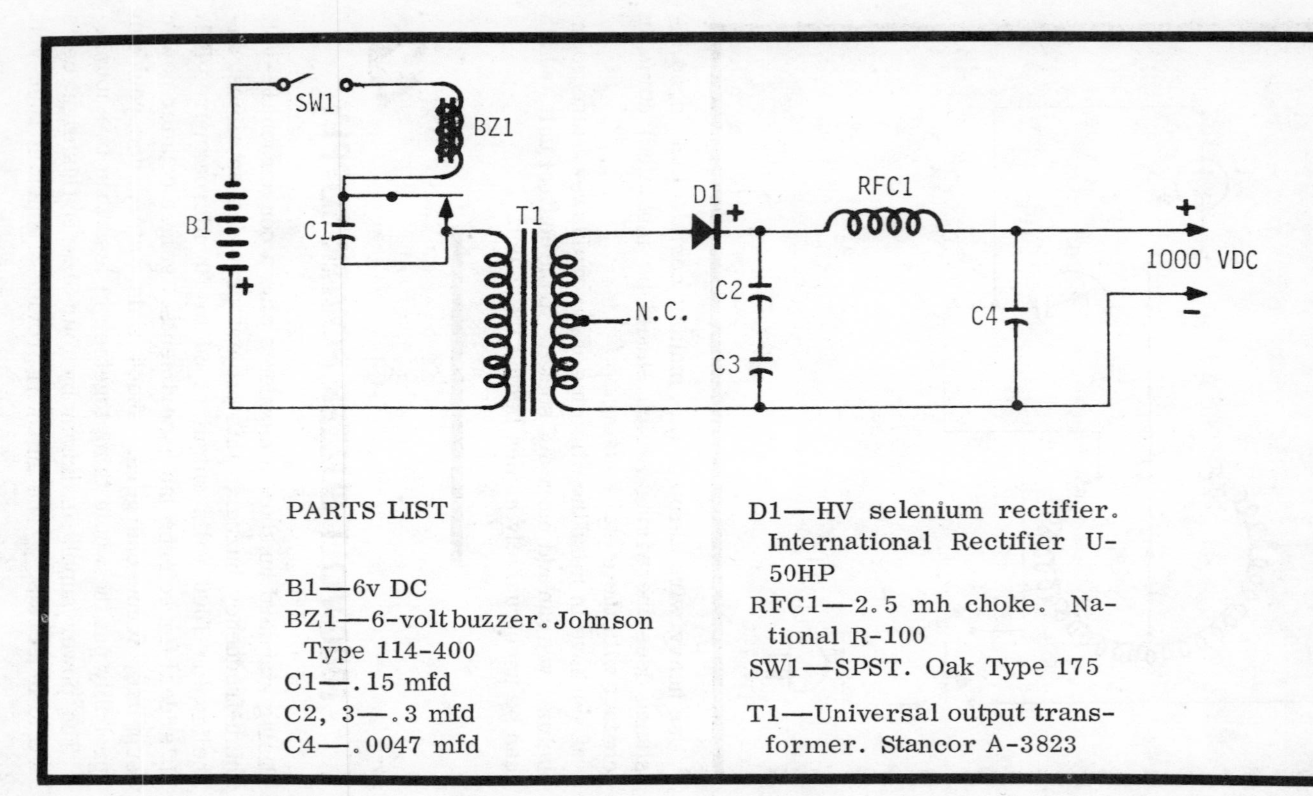

PARTS LIST

B1—6v DC
BZ1—6-volt buzzer. Johnson Type 114-400
C1—.15 mfd
C2, 3—.3 mfd
C4—.0047 mfd

D1—HV selenium rectifier. International Rectifier U-50HP
RFC1—2.5 mh choke. National R-100
SW1—SPST. Oak Type 175
T1—Universal output transformer. Stancor A-3823

output for 60 hours if left running continuously and 150 hours if you use it for only an average of 2 hours per day.

When you've finished construction, turn on SW1. Now, adjust the buzzer for the highest-pitched note while at the same time striving for highest possible output voltage.

A word of caution: This power supply is no plaything. It'll give one heck of a nasty sting. For this reason be sure that all contact points are well insulated and that it is well-housed in a protective case above ground.

43

BLOW-OUT PROTECTOR FOR LOUDSPEAKERS

Although it's often assumed that loudspeakers supposedly never blow out, if too much power is pumped into them they'll literally split at the seams. In normal use, of course, this seldom happens. But if you're a teenager who likes to "turn on" with full-blast music or a hi-fi addict who appreciates heavy volume, this gadget may save you a great deal of trouble.

Many people don't realize that the distortion they hear in old car radios and some hi-fi sets is not at all the fault of the final audio tube, but in fact the sad result of an over-driven (blown out) loudspeaker. Worse yet, the audio tones responsible for all the damage are frequently those out of range of the human ear. So, while you're pushing up the volume to hear the lead singer, you may be overdriving the speaker without realizing it.

The instrument shown in the accompanying schematic will signal, by the flashing of a neon bulb, when you are approaching this critical point if you program the thing correctly. First off, you must determine the actual power handling capability of the speaker.

For purposes of illustration, suppose your speaker is rated at 25 watts and is of the 16-ohm impedance variety. Using the same formula employed in Project 32, we can see, then, that this wattage represents 20 volts at the speaker voice coil

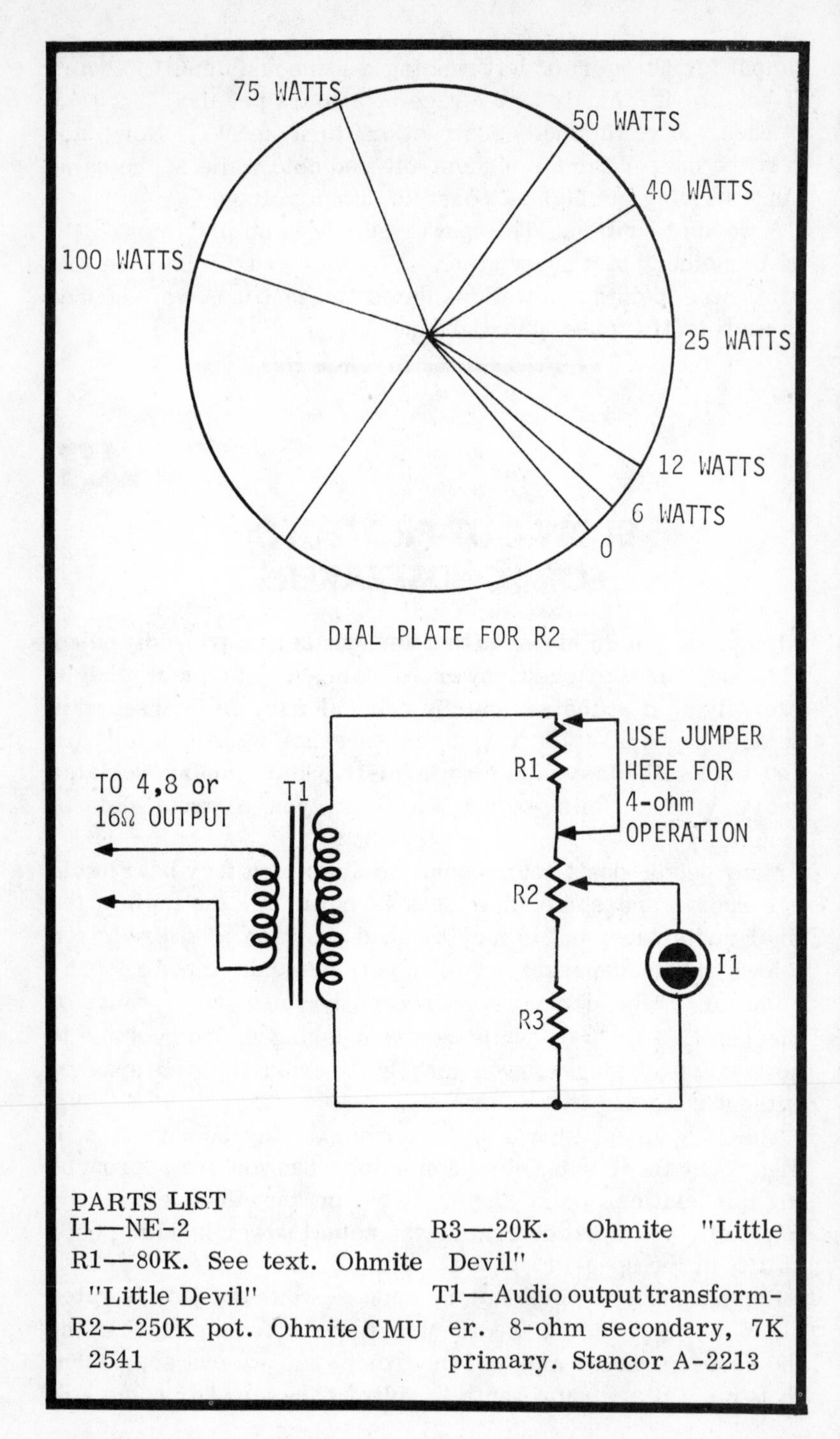

PARTS LIST

I1—NE-2

R1—80K. See text. Ohmite "Little Devil"

R2—250K pot. Ohmite CMU 2541

R3—20K. Ohmite "Little Devil"

T1—Audio output transformer. 8-ohm secondary, 7K primary. Stancor A-2213

terminals. By inserting the blow-out protector, the transformer steps up this voltage by a 30:1 ratio to 600 volts developed across R1, R2, and R3. Since the neon lamp requires only 65 volts to flash, we adjust R2 so that just a mite over 10% of this voltage is felt by the neon bulb.

Simplifying the procedure, you just set R2 to approximately 7 to 9K up from the 20K resistor end of the potentiometer so that the total resistance across the two is on the order of 25K. Other wattage settings are arrived at in the same manner.

For added accuracy, if you're using a 4-ohm speaker eliminate R1 altogether; if using an 8-ohm speaker, use a 51K resistor instead of the 80K suggested, or if you have a 16-ohm speaker, use a 121K resistor. If you want to incorporate the dialplate shown, use the resistive values just specified.

INEXPENSIVE TRANSISTOR CHECKER

Here's a very inexpensive transistor checker that will save you a lot of frustration and expense in your project building. It performs some basic checks which will rapidly tell you if a transistor is excessively leaky or if it has become internally short-circuited. Additionally, the checker will tell the approximate beta value, frequently referred to in catalogs and manufacturers' literature. In this way you can see how your transistors stack up against the spec sheets.

Essentially a very simple instrument, not only to build but to operate, the main control switch is more or less halved to permit testing of both PNP and NPN types. Make sure of the transistor type before inserting it in the checker.

For the most part, the lower the amount of leakage the better the transistor is. Understandably, however, the higher-priced transistors will register lower leakage than the 8-for-88¢ variety. To check gain (beta) switch to the "Gain" position and notice the difference in meter readings. If there seems to be an excessive amount, the transistor is probably shot.

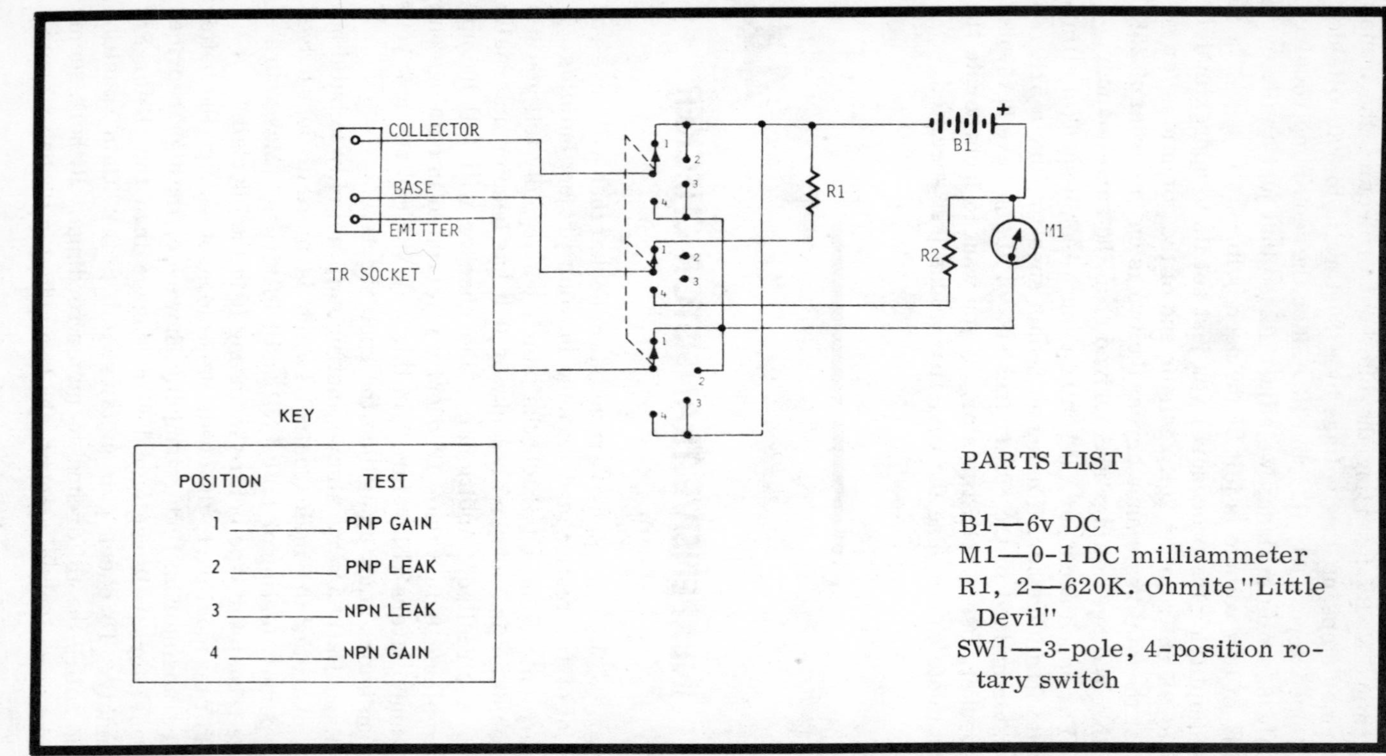

COLLECTOR
BASE
EMITTER
TR SOCKET
R1
R2
B1
M1
KEY
POSITION TEST
1 PNP GAIN
2 PNP LEAK
3 NPN LEAK
4 NPN GAIN
PARTS LIST
B1—6v DC
M1—0-1 DC milliammeter
R1, 2—620K. Ohmite "Little Devil"
SW1—3-pole, 4-position rotary switch

Here are some representative readings on common garden-variety transistors: For a 2N107, you should register 0.1 on "Leak," and 0.5 on "Gain"; for a 2N137, check for 0.1 on "Leak" and a 0.7 on "Gain"; for a CK722, look for a 0.1 "Leak" and 0.3 "Gain". The many substitutes will also fall fairly close to these readings.

45

FREE-POWER BATTERYLESS POWER SUPPLY

Anyone who's read earlier hobbyist project handbooks will remember the famous free-power radios which went on to fame in connection with electronic eavesdropping in "The Electronic Invasion" and later in magazine articles and TV shows. Although the concept has been blown way out of proportion, the simple fact is that it is possible to convert radio signals to rectified DC..

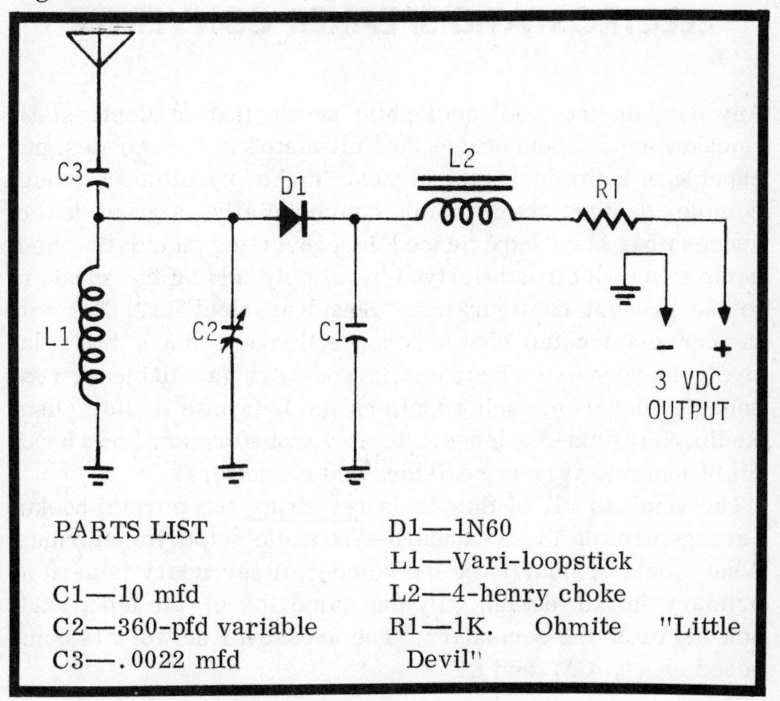

The simple circuit shown in the accompanying diagram, however, will provide enough power to drive a small transistorized amplifier, receiver, or even a tiny subminiature electric motor! During tests conducted off 20-mile-distant signals we've been able to obtain 3 volts at 300 microamperes, or 1 volt at nearly 1000 microamperes. If you live really close to a 50,000-watt radio transmitter, you may get results even better than this.

The circuit is essentially a radio receiver (crystal detector) tuned to the loudest signal on the band. For science fair work, a 0-500 DC microammeter tied across the output makes a dandy impression. Incidentally, don't forget to use a 0-5 volt voltmeter.

46

ELECTROSTATIC SPEAKER CONVERTER

Any dyed-in-the-wool audiophile knows that an electrostatic speaker is just about next to the "ultimate" in tweeter designs, capable of reproducing the highest "highs" possible. Although complex designs are available commercially, a great deal of success has been experienced in converting an existing hi-fi setup to an electrostatic type by simply adding the speakers to the present configuration. Sound unscientific? Not with the converter setup shown here. All you need are two relatively inexpensive electrostatic speakers (available at reasonable cost from such suppliers as Lafayette Radio, Olson Radio, Burstein-Applebee, etc.), a transformer, and a handful of junkbox-variety resistors and capacitors.

The trick to all of this is in reversing the normal hookup arrangement of T1, a standard-type audio output transformer. What would ordinarily be the voice-coil secondary is used as primary in our design. By the same token, the input leads now serve as the secondary. The crossover network is composed of R1, C1, and C2.

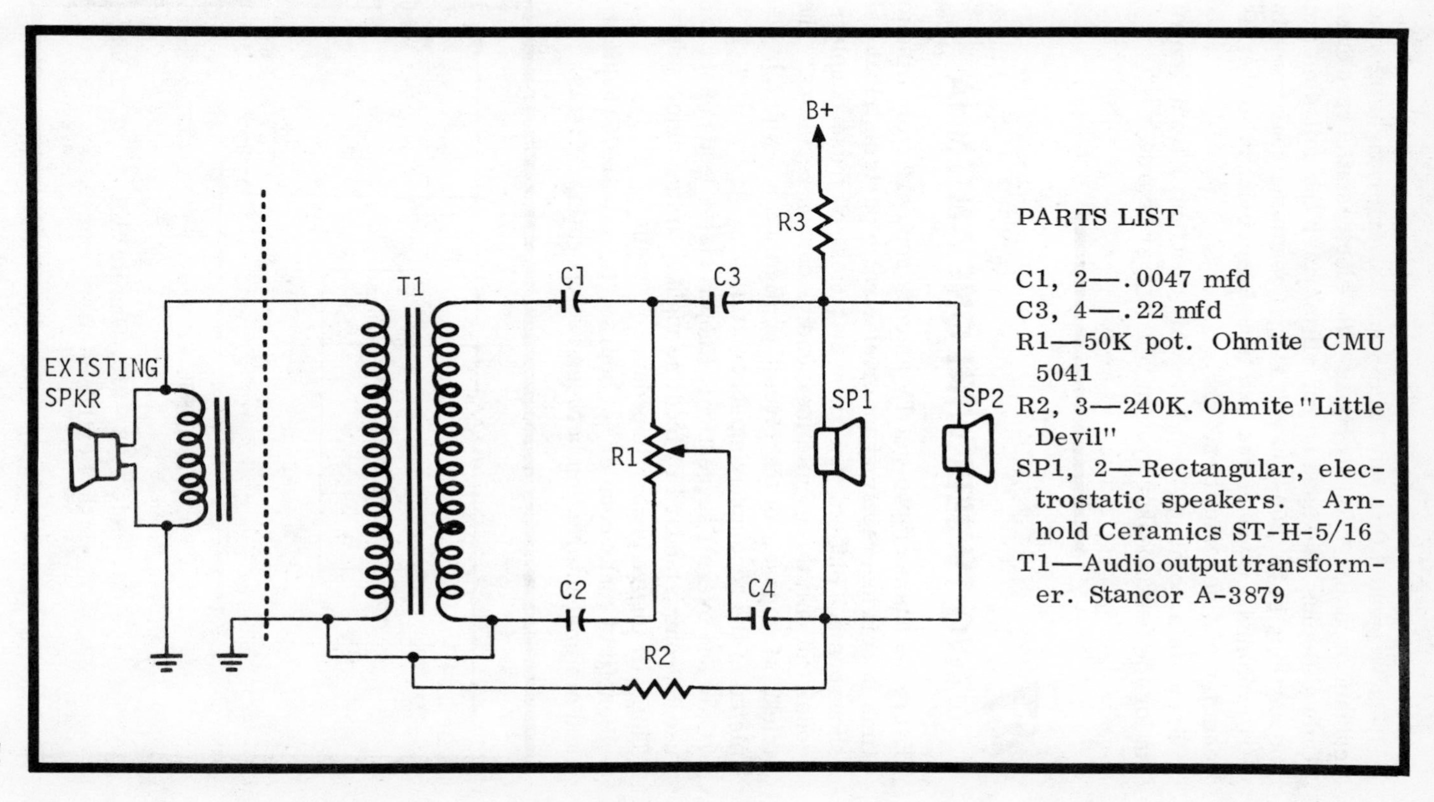
B+
R3
T1
C1
C3
EXISTING SPKR
SP1
SP2
R1
C2
C4
R2
PARTS LIST
C1, 2—.0047 mfd
C3, 4—.22 mfd
R1—50K pot. Ohmite CMU 5041
R2, 3—240K. Ohmite "Little Devil"
SP1, 2—Rectangular, electrostatic speakers. Arnhold Ceramics ST-H-5/16
T1—Audio output transformer. Stancor A-3879

It is suggested that you build your converter right into the cabinet or baffle used to house the electrostatic speakers. Only a B-plus line need be run off to power the unit (this can be anything in the 200-300 volt range) assuming your grounds are common. Connect the two transformer leads to the 16-ohm tap of your amplifier.

From this point on, you have only to adjust R1 and you're in for the best speaker system you've ever heard!

47

TWO TV SETS WITH ONE ANTENNA

Everyone has seen two-set TV antenna couplers from time to time in service-repair shops or electronic parts distributors' showrooms. But did you know that most of these couplers consist of only three quite inexpensive carbon resistors? The trick, of course, is the circuit configuration—which is depicted in the accompanying schematic.

Only one overall suggestion: Keep resistor leads short and use 300-ohm standard TV feedline right up to the solder joints. This will insure proper impedance matching.

Benefit of such a coupler, of course, is the isolation achieved and minimal interaction between two operating TV sets.

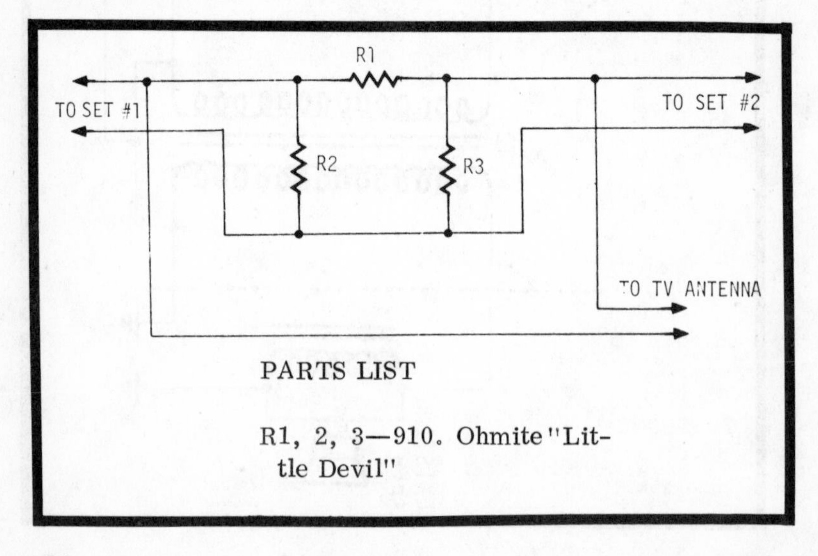

PARTS LIST

R1, 2, 3—910. Ohmite "Little Devil"

48

TUBE REJUVENATOR

Want to know the secret behind any of those "bargain tubes" you see in some of the leading electronics monthlies? Well, here it is: A little-known rejuvenator that does an amazing job of bringing "dead" tubes back to life. Why pay $1.00 apiece for standard radio and TV tubes when you can restore them yourself? Providing the tube isn't burned out altogether (most aren't), you'll be amazed at what this little gem can do!

In use, switch SW1 to the TEST position and see what happens. Neon bulb I1 will light immediately unless the filaments are burned or open. Discard any tubes that fail to register a light on I1. Next, take the tube (with the good filaments) into either position A, B, or C on SW1. This is determined by the tube's specifications, which can be found in any standard tube manual. The rejuvenator will work only with tubes containing 6.3v AC heaters (6AU6s, etc.), although if you add another 9-pin socket with pins 4 and 5 soldered together as one filament terminal and pin 9 as the other, you can charge up the 6.3/12.6-volt tubes such as 12AT7s, 12AX7s, etc.

Here's what to do: After consulting the manual, observe that SW1 position A provides .6 amp of current, position B provides .3 amp, and position C provides .15 amp. Merely select the proper current rating based on what you've found in the manual, and proceed.

Allow your tube to cook for 10 minutes in the rejuvenator before replacing in the original unit or checking with a good tube tester. If it's still weak, repeat the process. Although this unit's batting average is extremely high, there'll always be a few rotten eggs that just cannot be revived. In any case, follow the three-strikes-you're-out philosophy before discarding any tube that gets a good light on the neon bulb.

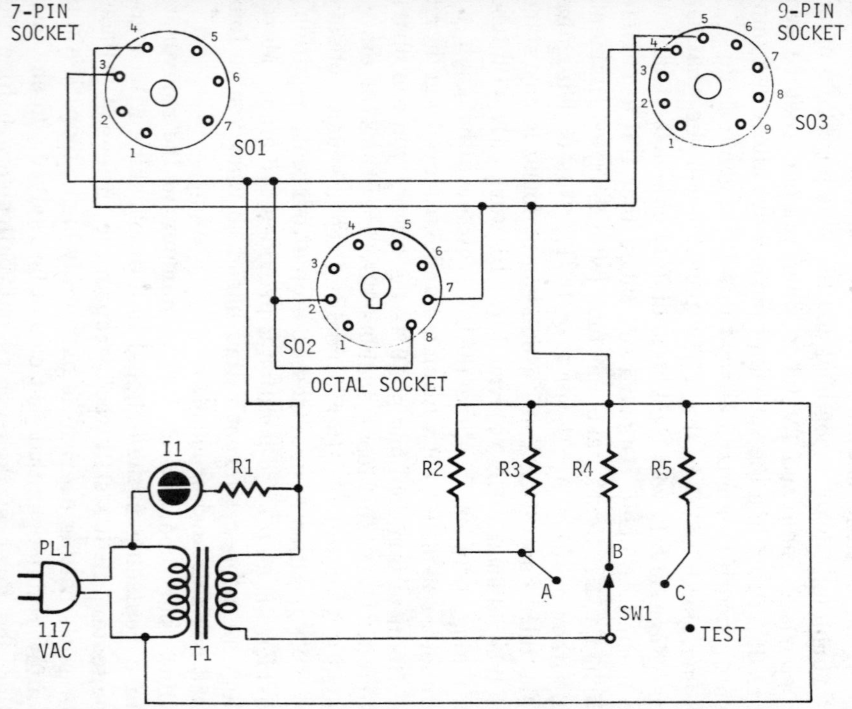

PARTS LIST

I1—NE-16
PL1—AC plug. Amphenol 61-M11
R1—62K. Ohmite "Little Devil"
R2, 3, 4—15, 2 watts. Ohmite "Little Devil"
R5—33, 2 watts. Ohmite "Little Devil"
SO1—7-pin socket. Amphenol 147-500
SO2—Octal socket. Amphenol 88-8X
SO3—9-pin socket. Amphenol 59-409
SW1—5-position rotary
T1—12v AC filament transformer. Stancor P-8130

49

OSCILLOSCOPE CALIBRATOR

Here's a gadget that can enhance the performance of your oscilloscope 200%, yet cost you next to nothing to put together. While oscilloscopes are great for telling you just about all you need to know about a device under test, suppose you want to see the actual peak voltage of a waveform closeup? Just hook our handy calibrator in between the unit and your scope and you've got it made.

How do you use it? Well, suppose you wanted to measure voltage being fed to the oscilloscope. First thing you'd do is switch SW1 to the CAL position and (for the sake of discussion) the PEAK-TO-PEAK control to the 10-volt setting. Adjust the scope to provide a specific number of "boxes" on the screen. If you like, adjust so that you get 10 boxes. Now you can see that 10 volts from the calibrator equals 10 boxes on the screen. Obviously, then, when you switch over on SW1 to the SCOPE position, you'll have a visual voltmeter reading in direct proportion. If, for example, you now see 15 boxes, you've got 15 volts going in. Suppose you adjusted the scope (when the calibrator was on CAL) so that you saw only one box. Then if you switched over to SCOPE and saw, for example, 6 boxes, you'd know you were feeding 60 volts in.

As with any precision device, our calibrator needs to be calibrated itself before it can be effectively used. Refer to the "Calibrator Calibrator" circuit in the accompanying diagram. The idea behind this thing is to feed 17.8 volts into the calibrator, since this corresponds to a peak-to-peak voltage of 50 volts which we use to calibrate the oscilloscope.

Once again get 10 boxes on the oscilloscope screen with SW1 in the SCOPE position. Now switch over SW1 to the CAL po-

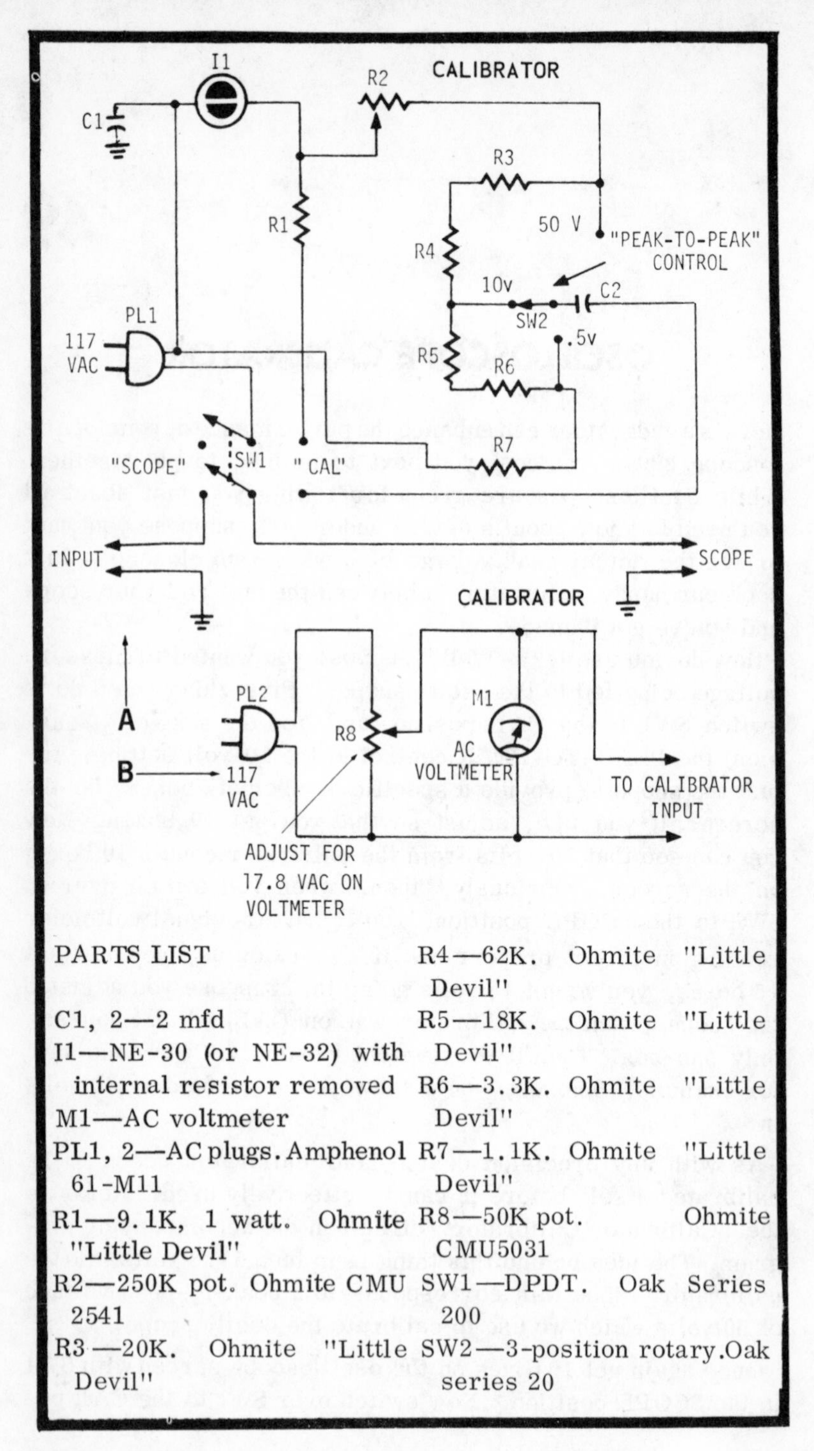

PARTS LIST

C1, 2—2 mfd
I1—NE-30 (or NE-32) with internal resistor removed
M1—AC voltmeter
PL1, 2—AC plugs. Amphenol 61-M11
R1—9.1K, 1 watt. Ohmite "Little Devil"
R2—250K pot. Ohmite CMU 2541
R3—20K. Ohmite "Little Devil"
R4—62K. Ohmite "Little Devil"
R5—18K. Ohmite "Little Devil"
R6—3.3K. Ohmite "Little Devil"
R7—1.1K. Ohmite "Little Devil"
R8—50K pot. Ohmite CMU5031
SW1—DPDT. Oak Series 200
SW2—3-position rotary. Oak series 20

sition with the PEAK-TO-PEAK control in the 50-volt setting. Now adjust R2 until you get the same ten boxes on the oscilloscope screen. To check your settings, switch the PEAK-TO-PEAK switch to the 10-volt position without touching anything else. Now you should see two boxes on the oscilloscope.

---

50 SEQUENTIAL NEON FLASHER

Have you seen some of the high-style ripple effects the automakers are producing in rear-end taillights? The desirable sequential flasher they use operates on a principle similar to that shown in the schematic here, only this project is sort of a scaled-down model of the Big Three's.

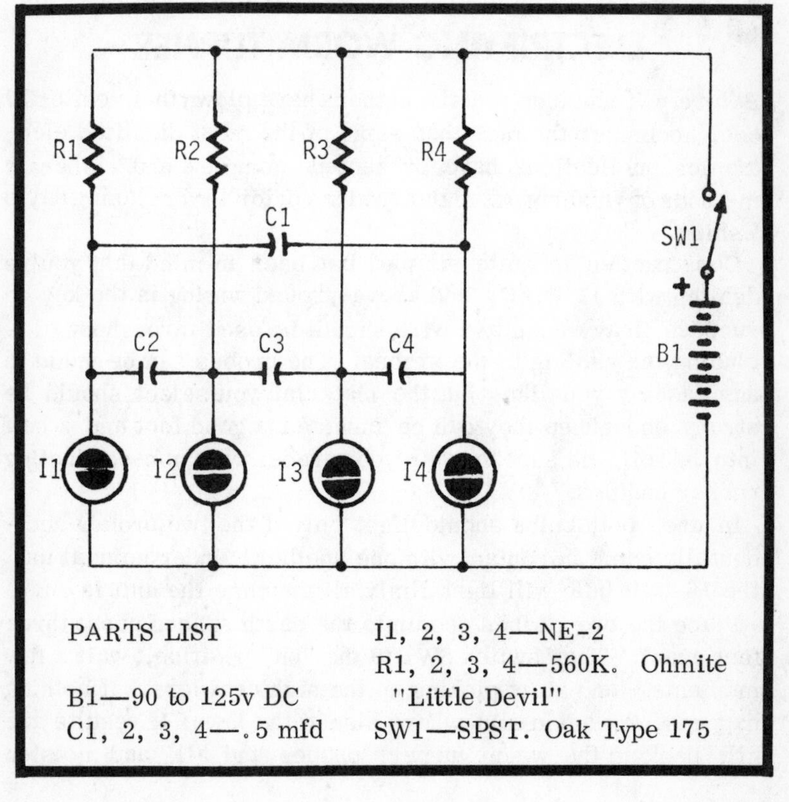

PARTS LIST

B1—90 to 125v DC
C1, 2, 3, 4—.5 mfd
I1, 2, 3, 4—NE-2
R1, 2, 3, 4—560K. Ohmite "Little Devil"
SW1—SPST. Oak Type 175

You can make attractive displays from the arrangement shown here, providing all bulbs are good and the battery supply is in fair shape. Incidentally, you can string a series of cheap 9-volt TR batteries to obtain the voltage necessary for this project, add a couple of 67 1/2-volt "B" cells together, or put together the inexpensive voltage quadrupler shown in Project 102. In any case, it makes a dandy amusement device for the kiddies and a perpetual conversation piece for the adults.

By the way, if you like, install a 1-meg potentiometer across the plus and minus wires leading from the batteries. In this manner you can, to some extent, regulate the frequency of the flashes, slowing them down and speeding them up at will. Regardless, however, the bulbs will always go on and off in sequence.

---

51 ELECTRIFYING WORM TURNER

Before you conclude that the authors have blown their editorial cool, consider the fact that some of the most dignified electronics publications have presented complex and elaborate methods of flushing out nightcrawlers prior to a relaxing day's fishing.

Construction is quite simple, but bear in mind that you're dealing with 117v AC, and above-ground wiring is the key to success. Heavy insulated wire should be used throughout, including the cabling to the probes. The probes can be made in any fashion you like, but the material you select should be strong enough so they can be inserted a good foot and a half into the soil. Be sure to equip your probes with non-conducting rubber handles.

In use, both bulbs should light only if the two probes accidentally come in contact with one another. Under normal use, the 15-watt bulb will light dimly all the time the unit is on.

Place the two probes deep into the earth spaced about three feet apart. Then switch SW1 to the "on" position, wait a few moments, and start picking up the nightcrawlers. If nothing happens, try again somewhere else in the lawn. If results are nil, pack up the worm turner, probes and all, and moisten the earth with a good hosing. Then try again. Only in extremely dry weather should you have to hose down the soil.

CAUTION: Never attempt to "plant" or move the probes when the worm turner is on.

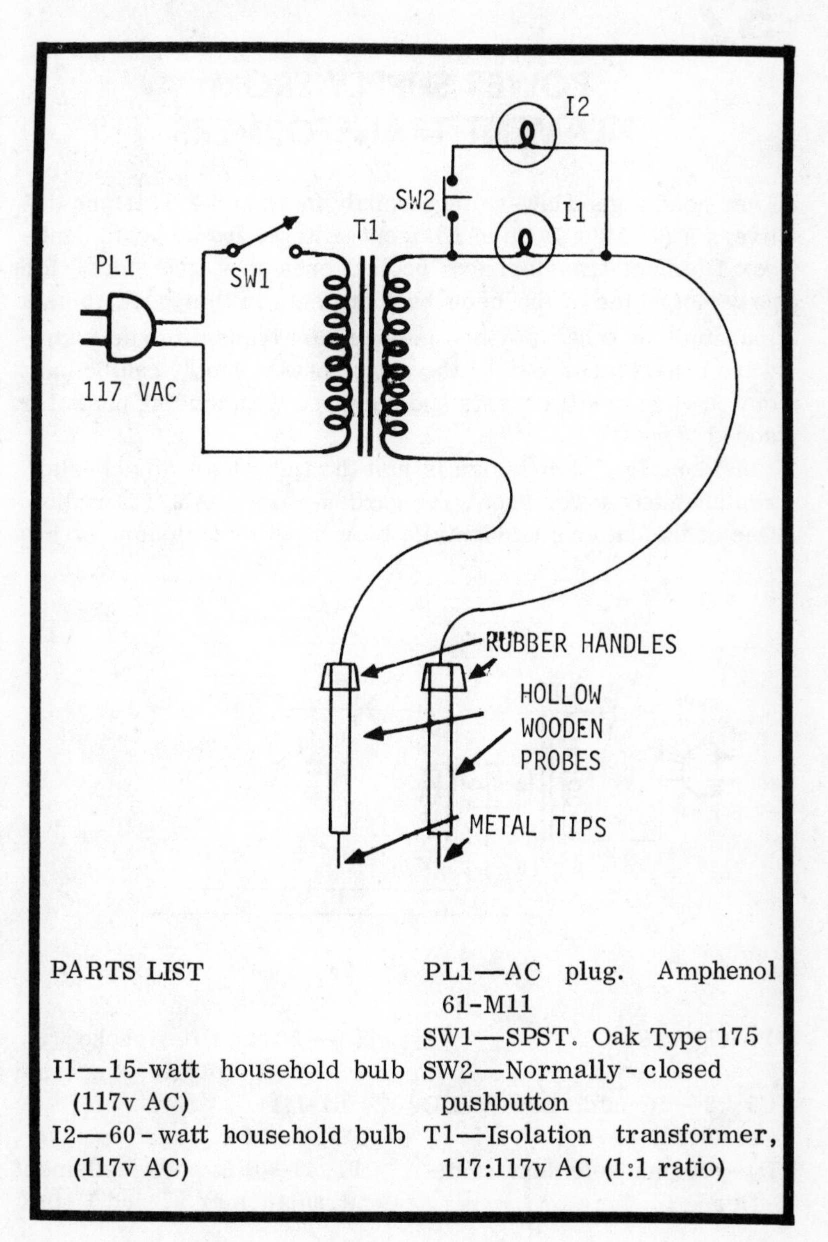

PARTS LIST

I1—15-watt household bulb (117v AC)

I2—60-watt household bulb (117v AC)

PL1—AC plug. Amphenol 61-M11

SW1—SPST. Oak Type 175

SW2—Normally-closed pushbutton

T1—Isolation transformer, 117:117v AC (1:1 ratio)

52

POWER SUPPLY FROM FILAMENT TRANSFORMERS

Ever need a good low-voltage supply in a pinch? This one delivers a full 120v DC plus filament voltages from cheap, junk-box filament transformers hooked back-to-back. Ideal for powering some of the neon bulb projects in this book, plus a multitude of other devices (novice ham transmitters, short-wave converters, etc.), the entire power supply can be built onto a very small chassis and equipped with binding posts for added versatility.

Incidentally, this design is just the ticket as a "quick-slap" replacement power supply for garden-variety AM/FM radios. One of the authors accidentally blew a power transformer in a

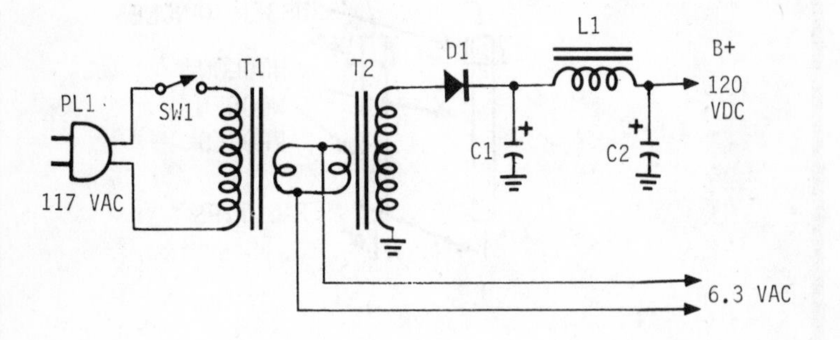

PARTS LIST

C1, 2—16 mfd, 200 WVDC electrolytic
D1—70 ma selenium rectifier
L1—50 ma filter choke
PL1—AC plug. Amphenol 61-M11
SW1—SPST. Oak Type 175
T1, 2—6.3v AC filament transformer. Triad F-14X

Japanese FM set and simply hooked this power supply in place of the original. True, B-plus is a bit lower than the circuit called for, but you'll never convince the most descriminating listener of any changes. By the way, the entire power supply fit neatly inside the receiver.

53

COLOR TV PICTURE ROLL STOPPER

If you live within a reasonable distance from a television station, chances are you often are receiving a signal far in excess of what the tuner in your TV is capable of handling. And, with the advent of color television antennas—which, for the most part, do an excellent job of increasing overall gain—the result is frequently that your picture may darken and roll to the right of the screen between station breaks or whenever the picture contrast control is advanced.

One solution is to construct the roll stopper shown in the accompanying schematic diagram. Resistance values chosen should be determined by the degree of attenuation (in decibels) you wish to achieve. If you're unsure, you can experiment with potentiometers in place of the fixed composition resistors shown until you get a handle on your requirements.

Generally, however, here are our recommendations for values, starting at minimum attenuation: 1) for 5 db attenuation, use 47-ohm resistors for R1, 2, 4, and 5 and a 510-ohm value for R3; 2) for 10 db attenuation, use 82 ohms for R1, 2, 4, and 5 and a 220-ohm resistor for R3; 3) for 15 db attenuation use 120 ohms for all resistors; 4) for 20 db attenuation use 120 ohms for R1, 2, 4, and 5 and 62 ohms for R3: 5) for 25 db attenuation use 150 ohms for R1, 2, 4, and 5 with a 36-ohm resistor for R3; 6) for 30 db attenuation use 150 ohms for R1, 2, 4, and 5 in conjunction with a 22-ohm resistor for R3.

Bring your 300-ohm feedline leads directly to the roll stopper interconnection joints for the best overall performance.

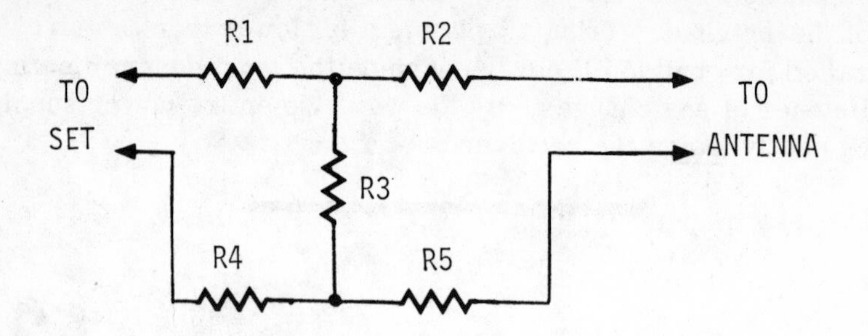

PARTS LIST

R1, 2, 4, 5—47, 82, 120 or 150 ohms, Ohmite "Little Devil." See text

R3—22, 36, 62, 120, 200 or 510 ohms, Ohmite "Little Devil." See text

54

ELECTRONIC ANEMOMETER

Normally when one thinks about an anemoneter, a wooden-cross affair comes to mind, whirling around like some kind of a horizontal windmill. This version, however, employs no moving parts, yet permits reading wind velocity from a meter with amazing accuracy. Actually, a combination of an electronic thermometer and hygrometer, the electronic anemometer uses two rugged thermistors to "sense" the wind on

PARTS LIST

B1—4v DC

M1—0-2 DC milliammeter

R1, 5—240. Ohmite "Little Devil"

R2—100-ohm pot. Ohmite CMU-1011

R3—3.3K. Ohmite "Little Devil"

R4—1K pot. Ohmite CMU 1021

R6—510. Ohmite "Little Devil"

R7, 8—11. Ohmite "Little Devil"

R9—500-ohm pot

SW1—DPDT pushbutton

SW2—Pushbutton, normally-open

TH1—Matched thermistor pair. Victor Engineering Corp., Type A-33 (25° C types) or equivalent

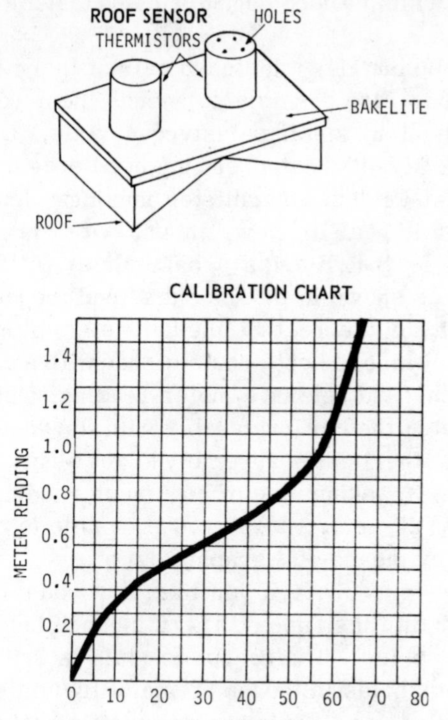
ROOF SENSOR
HOLES
THERMISTORS
BAKELITE
ROOF
CALIBRATION CHART
METER READING
1.4
1.2
1.0
0.8
0.6
0.4
0.2
0
10
20
30
40
50
60
70
80

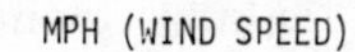
MPH (WIND SPEED)

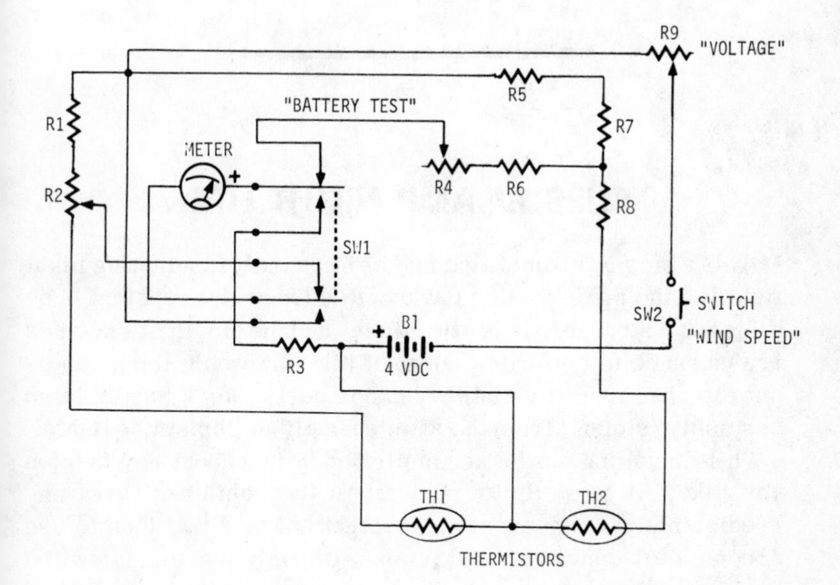
R9
"VOLTAGE"
R5
"BATTERY TEST"
R1
R7
METER
R4
R6
R2
R8
SW1
SWITCH
SW2
"WIND SPEED"
B1
R3
4 VDC
TH1
TH2
THERMISTORS

the roof and a simple decoding unit inside that interprets what's happening.

Although you can buy a matched pair of thermistors already mounted, it is also possible to mount them yourself by repackaging them in small polystyrene vials. Drill (or better yet use a "red hot" heated sewing needle) four tiny holes in one of the vial-enclosed thermistor housings, leaving the other vial completely sealed. Now, mount both vials on a piece of polystyrene or bakelite (as a base plate) with glue. Do not drill. Wire as shown in the diagram, and the job's half done.

The circuit for the rest of the unit is simple, but requires calibration, which is best accomplished with a controlled wind that can be checked against a meter reading. One way of doing this is to get in the car (preferably with someone else driving) and allow the thermistor assembly to get plenty of wind as you drive along. By taking note of how much reading you get at 5, 10, 15, 20, 25, etc., MPH you'll be able to calibrate your instrument by means of a graphic curve.

You can use our chart if you like, but you'd better check it against your own instrument's performance before you accept it strictly "as is." The reason for this is that you may have made the four holes in the top of one of the vials smaller than we did. In that case, enlarge them sufficiently so that, say, a 60 MPH speed corresponds to a 1 ma reading on your meter.

55

BUBBLE LAMP NIGHT LITE

Here's a simple project that makes for a truly fascinating night light for the kiddies. In a darkened bedroom the constant bubbling has a strangely hypnotic effect, and the dim light provides reassurance to children. Best of all, the whole thing can be put together from available junkbox parts plus a bubble lamp assembly "stolen" from the attic's supply of Christmas lights.

While mounting can be accomplished in just about any fashion you like, it is well to remember that children are unpredictable. Consequently, it might not be a bad idea to use a completely enclosed metal box with only the toggle switch

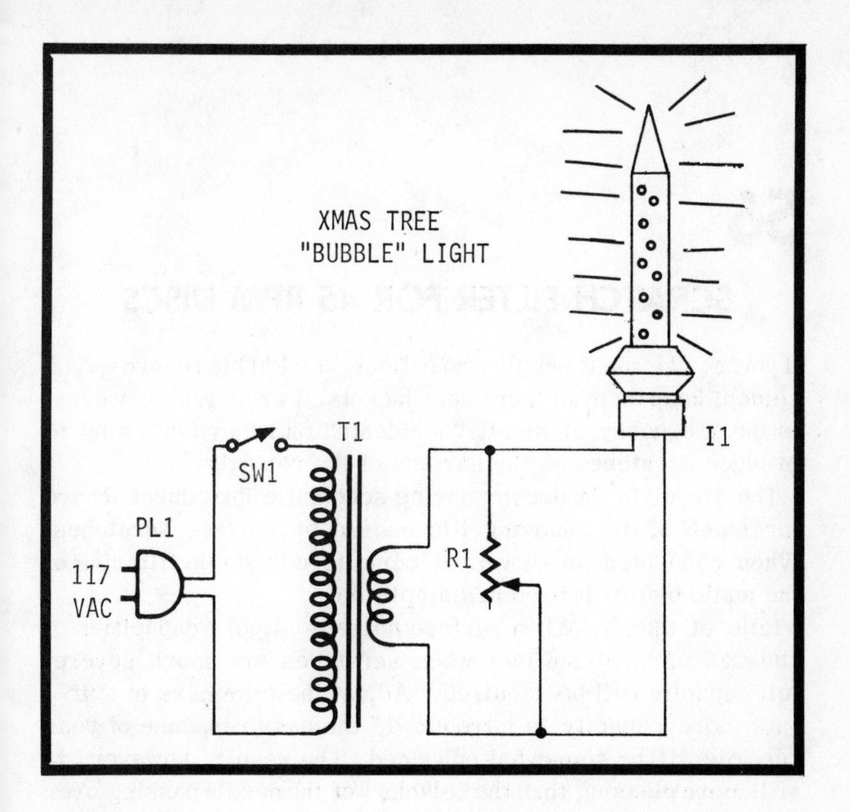

PARTS LIST

I1—Christmas tree bubble lamp (and socket)

R1—1-meg pot. Ohmite CMU 1052

PL1—AC plug. Amphenol 61-M11

SW1—SPST. Oak Type 175

T1—6.3v AC filament transformer, under 1-amp rating. Triad F-13X

and bubble lamp protruding. All wiring must be above-ground and well insulated, and the 117v AC line cord should pass through a rubber grommet going into the unit. Additionally, it would be a good idea to tie the cord in a solid knot just as it comes in through the grommet as a precautionary measure.

Bear in mind that the bubble lamp works on the heat principle. If you have trouble getting it to bubble, screw the bulb more firmly into the socket and check that the bubble tube is tight in place against the bulb. Frequently replacing the bulb works wonders. Lastly, R1 can be installed inside or outside the housing, or omitted altogether. It serves only to dim the bulb.

56

SCRATCH FILTER FOR 45 RPM DISCS

If you're like most people who collect 45 RPM hit records, you seldom keep them in their dust jackets. Far worse, however, is the frequency of small "accidents" that inevitably tend to produce scratches on the surface of the records.

The circuit in the accompanying schematic introduces a cure for much of the annoying ills caused by surface scratches. When connected as shown, it permits adjustable filtering of the audio before it becomes amplified.

Rule of thumb: When surface noise is slight, capacities of 100-225 pfd will suffice; when scratches are more severe, full capacity will be required. Adjust the trimmers to suit.

As more capacity is introduced, the bass response of your records will be somewhat enhanced. The result, however, is still more pleasing than the "clacks" of the needle passing over a gouge in the surface of the disc.

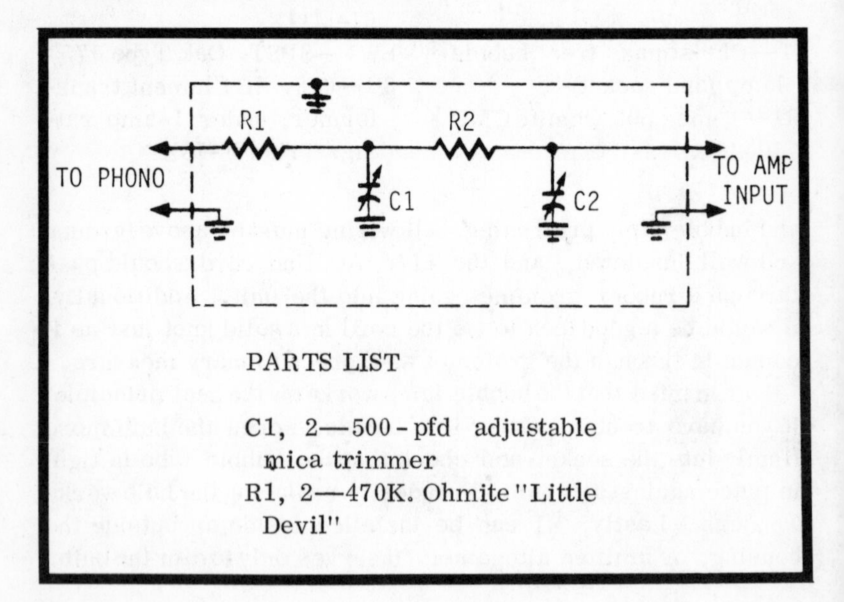

PARTS LIST

C1, 2—500-pfd adjustable mica trimmer
R1, 2—470K. Ohmite "Little Devil"

57

SUBMINIATURE POWER SUPPLY

Looking for a good source of low-voltage DC (9 to 12 volts)? Well, you've found it. This excellent little power supply can be used for any number of purposes, such as to power some of the projects in this book, as a supply for diode voltage triplers and quadruplers, or what-have-you.

Perhaps the primary governing factor in this project's size (it can be built into a 2 1/2 x 2 x 1 1/4" plastic case) is the fact that it uses a standard transistor audio transformer for the main power conversion. But don't take just any audio transformer. The one you use must be capable of handling 117v AC, the iron core must not run into saturation, primary impedance must be high enough at 60 Hz to keep current at a minimum under no-load conditions, and the primary-to-secondary ratio must be sufficient to furnish a reasonable amount of voltage when used for the purpose intended. We found that several Argonne types meet these requirements perfectly.

For general, all-around transistor project building this little power supply comes in mighty handy!

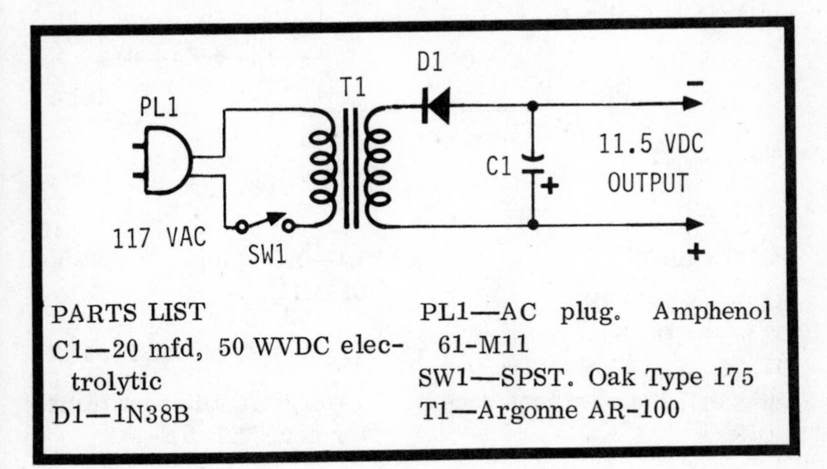

PARTS LIST

C1—20 mfd, 50 WVDC electrolytic

D1—1N38B

PL1—AC plug. Amphenol 61-M11

SW1—SPST. Oak Type 175

T1—Argonne AR-100

58

THE MYSTERY ALARM

Here's a circuit which will baffle your electronics-oriented friends, amuse the kiddies, yet provide a good time for all. Although some so-called experts may tell you that this thing can't possibly work, we suggest you build it in a transparent case so that the wiring is clearly visible and watch their jaws drop as you throw the switches.

In operation, the mysterious alarm works exactly the reverse of what you'd assume after studying the attached sche-

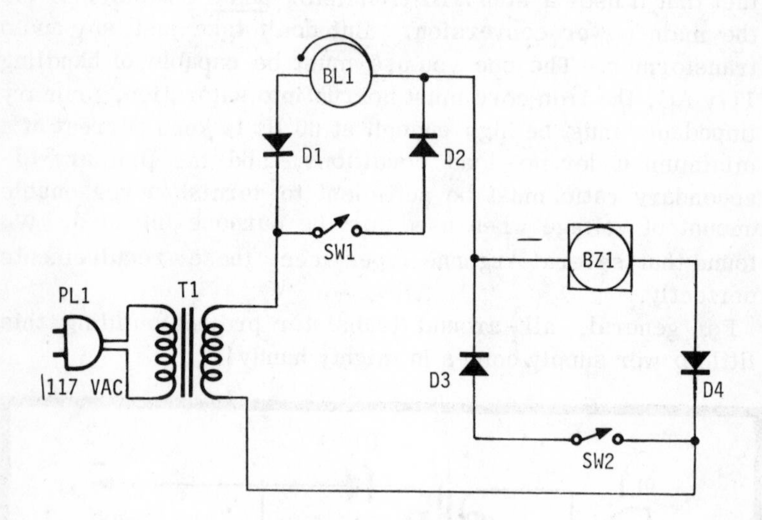

PARTS LIST

BL1—6v DC bell
BZ1—6v DC buzzer
D1, 2, 3, 4—15 PIV (or higher), 1-amp silicon rectifiers

PL1—AC plug. Amphenol 61-M11

SW1, 2—SPST. Oak Type 175

T1—6.3v AC filament transformer. Triad F-14X

matic. If you want the buzzer (BZ1) to sound, you throw switch SW1 (near the bell). Conversely, if you want the bell (BL1) to ring, you throw switch SW2. Impossible? Try it and see.

Standard junkbox parts are the order of the day. The only thing to watch for is that diodes D1 through D4 be at least of the 200 PIV type and be capable of handling at least 1 amp of current each. If you like, you can use subminiature International Rectifier Type 804s.

For more fun, build this in three sections, each in a clear plastic case connected together exactly as shown here. They'll never figure it out!

59

JUNKBOX MODULATION INDICATOR

Here's a dirt-cheap modulation indicator that instantly checks both the quality and percentage of audio you're pumping into your ham or CB rig. It even flashes a warning light whenever distortion or overmodulation is reaching your final amplifier! You can also monitor your modulation on a pair of headsets if you'd like to hear exactly what you sound like at the other end.

For best results, build this gadget into a separate Minibox connected by cables to your transmitter (it can be housed quite well in a 5 1/4 x 2 1/2 x 2 1/2" utility box).

Mount a jack on the transmitter to accommodate the insertion of the modulation indicator's three-wire cable, taking pains that the suggested in-transmitter circuit shown here is carefully added to the transmitter exactly as in the accompanying diagram. The value of R1 depends on the power of your transmitter (see chart).

Notice that .7 ma on your indicator meter represents exactly 100% modulation. By the same token, other readings show percentages of this in direct proportion. A .28 reading, for example, represents 40% modulation, while a .56 indicates 80%.

In use, throw the transmitter on with the indicator coupled in and the modulation control all the way off. Now push SW1

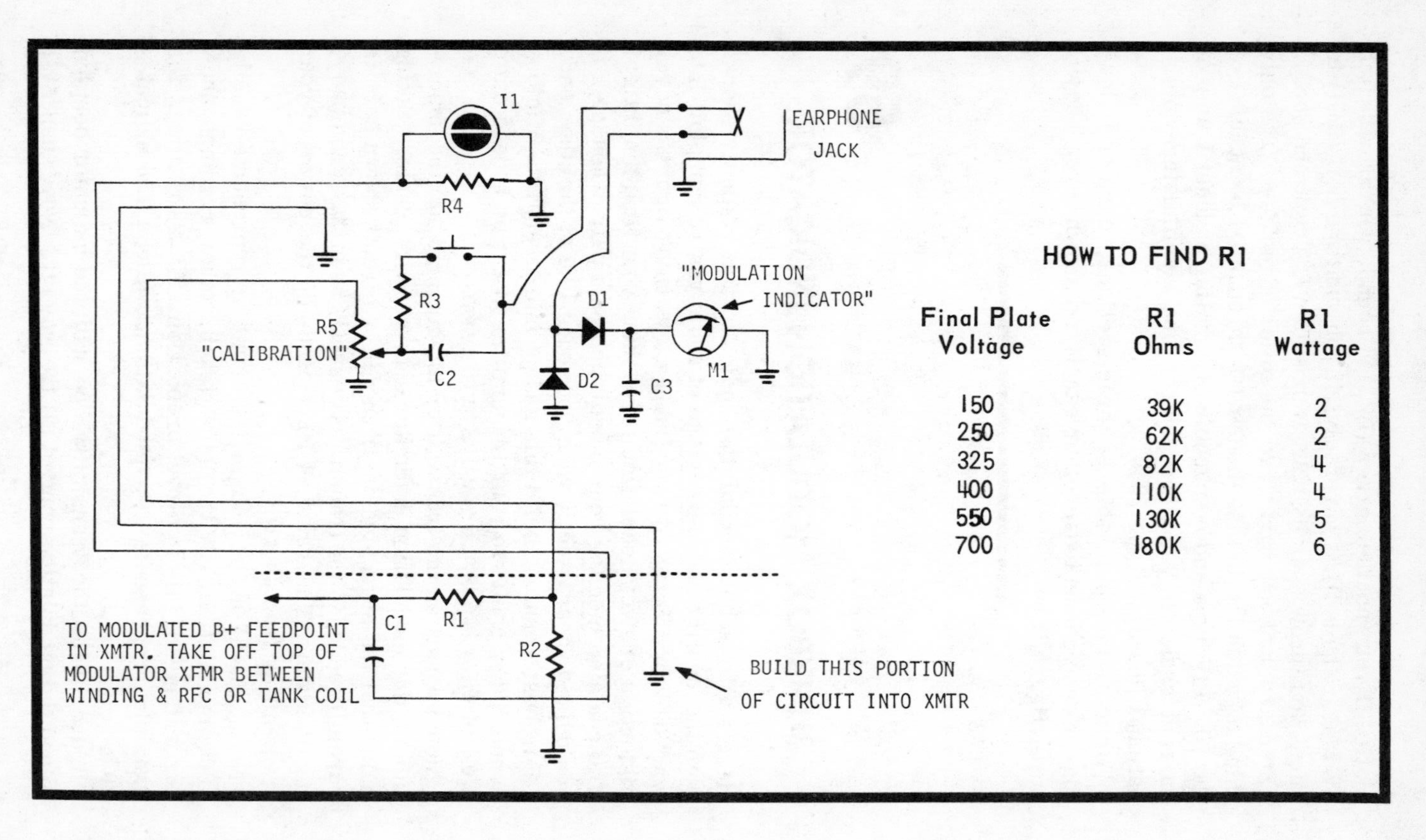

HOW TO FIND R1

Final Plate Voltage	R1 Ohms	R1 Wattage
150	39K	2
250	62K	2
325	82K	4
400	110K	4
550	130K	5
700	180K	6

PARTS LIST

C1—70 pfd
C2—10 mfd, 25 WVDC electrolytic
C3—110 mfd, 15 WVDC electrolytic
D1, 2—1N2070 silicon rectifiers
I1—NE-2A
M1—0-1 DC milliammeter
R1—See chart
R2—3.6K. Ohmite "Little Devil"
R3—7.5K. Ohmite "Little Devil"
R4—1.1 meg. Ohmite "Little Devil"
R5—10K pot. Ohmite CMU 1041
SW1—Pushbutton, normally-open

and simultaneously adjust R5 for a 1 ma meter reading. Release SW1 and start talking into the microphone while advancing the modulation control on your transmitter. Keep talking until you reach .7 ma on the meter (100% modulation). Now to check overmodulation, whistle into the mike and watch I1 flicker. Finally, plug in the headset and see what you sound like!

60

DARKROOM TIMER CONTROL

If you've ever done any photographic work, you can appreciate that your success ultimately depends on near split-second timing. In the darkroom prints can be ruined unless the proper exposure time is adhered to; hence, the multitude of mechanical and electrical timing devices currently available (some at pretty steep prices) in the camera shops. This project, though, consists of a reliable timer you can make with junkbox parts that will provide accurate timing for all printing and enlarging requirements.

An extremely simple item to build, the only point worth elaborating upon is the calibration. Using a clock with a second-hand (in a dimly-lit room) adjust R4 slowly until I1 is blinking at exactly one-second pulses. Now adjust R5 until I2 is blinking at 5-second intervals. Finish up by slowing tuning both R4 and R5 (ever so slightly) until the bulbs are in sync with each other.

For proper timing, you need only count the number of 1-second and 5-second pulsed flashes. For 23 seconds, for example, watch for four flashes of I2 and three of I1.

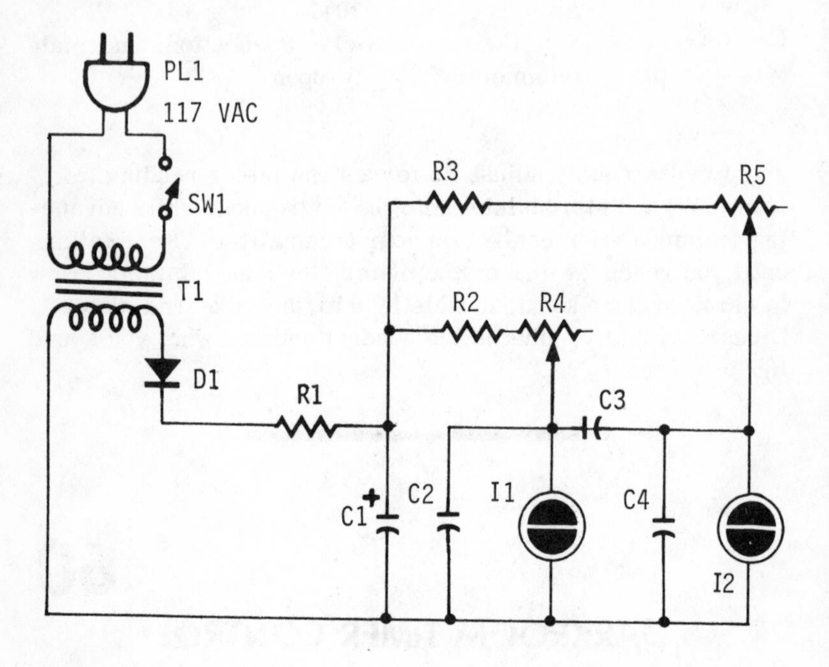

PARTS LIST

C1—10 mfd, 250 WVDC electrolytic
C2—.11 mfd
C3—.022 mfd
C4—.5 mfd
D1—25 ma selenium rectifier
I1, 2—NE-2
PL1—AC plug. Amphenol 61-M11
R1—1.1K. Ohmite "Little Devil"
R2, 3—18 meg. Ohmite "Little Devil"
R4, 5—5 meg. Ohmite "Little Devil"
SW1—SPST. Oak Type 175
T1—1:1 ratio, 117v AC isolation transformer

61

EFFECTIVE 6-METER ANTENNA COUPLER

Though the ham books and amateur radio magazines seem to abound with an impressive array of antenna couplers and matchers for the low-frequency bands (to 10 meters), when it comes to 50 MHz someone seems to have forgotten the lowly technician. The sad truth is that because of this, many final tubes are burned because of high SWR, and manufacturers either have to ignore this factor altogether or add it into their transmitters at the factory, consequently adding to the final price tag. As you can imagine the breakdown is something less than 50/50.

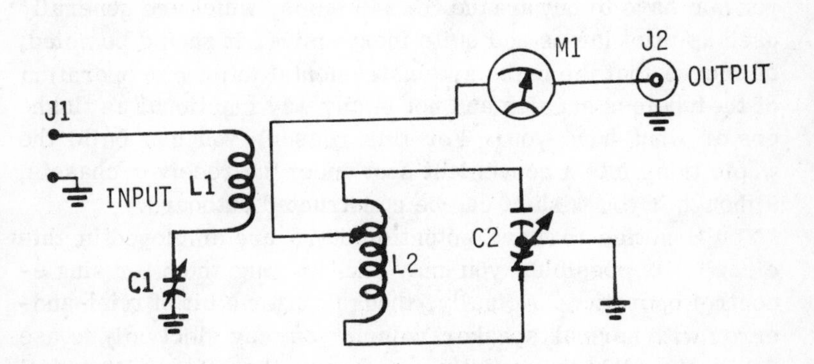

PARTS LIST

C1—150-pfd variable

C2—40-pfd, two sections, variable

J1, 2—SO-239 coax receptacles. Amphenol 83-1RTY

L1—3 turns of #16 wire, 1" diameter. 16 turns per inch. B & W 3015

L2—7 turns of #14 wire, 1-3/4" diameter. 4 turns per inch. Tapped 2 turns from one end. B & W 3021

M1—0-2 amp RF ammeter. Weston 507

Strick adherence to the specifications shown here, however, can cure your matching problems once and for all. Not one of those "all-around VHF couplers," this one's been designed expressly for the 6-meter enthusiast who's looking to get the utmost out of his system.

Hooked to a conventional 52-ohm line, tune-up is accomplished as follows: Using a good SWR bridge (Heathkits are excellent for this purpose), adjust the transmitter final and coupler for maximum meter reading on the bridge. Now, switch to reflected power (SWR) and tune slowly for minimum. Repeat this several times, and you'll be amazed how low your SWR will drop!

62

TUBELESS AUDIO SQUELCH

Here's a dandy receiver squelch you can throw together in no time with a minimum investment of parts. The only things you may have to buy are the GE #49 bulbs, which are generally used as pilot lamps and quite inexpensive. It should be noted, of course, that the bulbs are instrumental to proper operation of the tubeless squelch and not in any way functional as flashers or what-have-you. For this reason, you can build the whole thing into a convenient spot under the receiver chassis, although if you wish it can be constructed outboard.

You'll notice that two potentiometers are employed in this circuit. If possible, you may like to gang them for single-control operation. Actually, though, after a bit of trial-and-error with normal speaker volume you may elect only to use R2, setting R1 to a predetermined point that affords R2 a "full swing" for your particular receiver.

If your receiver's output speaker is something other than a 4-ohm type (which it probably isn't), you may have to install a matching transformer for maximum efficiency and performance of the audio squelch.

By the way, this thing is great for 2-meter enthusiasts lacking a BFO for catching those weak stations at the low end. Just adjust R2 for a slight above-the-noise-level break-in point, and you're in business!

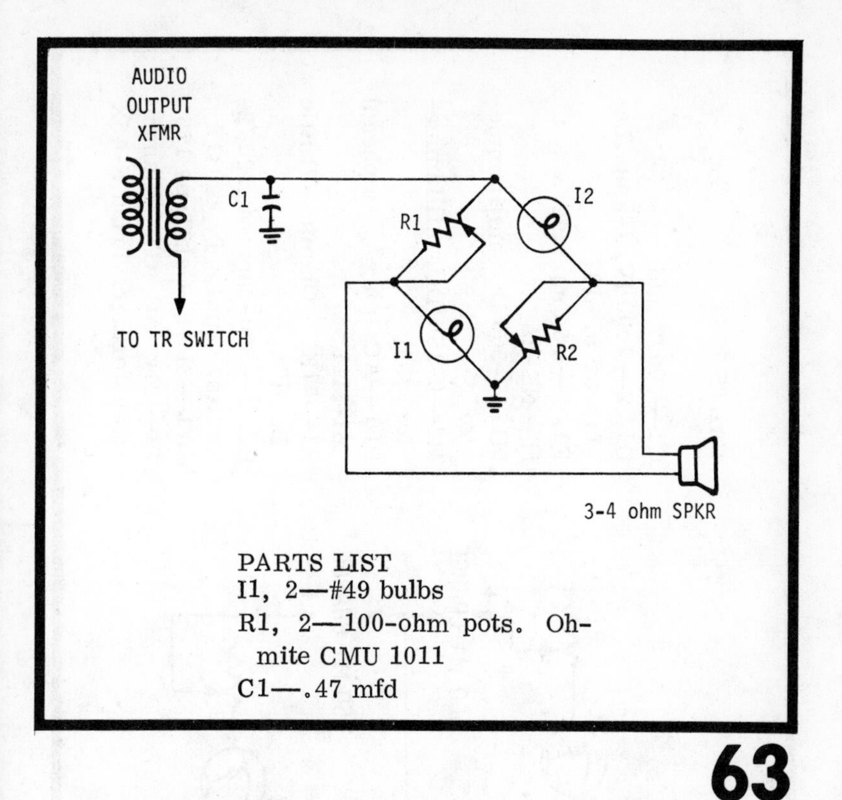

PARTS LIST
I1, 2—#49 bulbs
R1, 2—100-ohm pots. Oh-
mite CMU 1011
C1—.47 mfd

63
ALL-PURPOSE BATTERY CHARGER

Here's a battery charger unlike any other you will find in this book or elsewhere. Not only does it do an excellent job on any number of batteries (from 1 1/2 to 9v DC) commonly used in transistor circuits, but it also has been designed specifically with nickel-cadmium rechargeable batteries in mind.

A glance at the schematic will reveal that two separate output levels are provided: one for "regular charge" (the 150 ma contact points), and one for "super charge" (the 500 ma connector). Care should be exercised in selecting which you wish to use, since too many 500 ma whallops may ultimately shorten the recharging lifespan of your cells.

If you're using ni-cads, check the manufacturer's sheets before inserting the batteries. You'll notice that for best performance the ni-cads are supposed to become completely discharged at least four times per year. Additionally, the charging rate is specified clearly, based on a 15-hour recharging

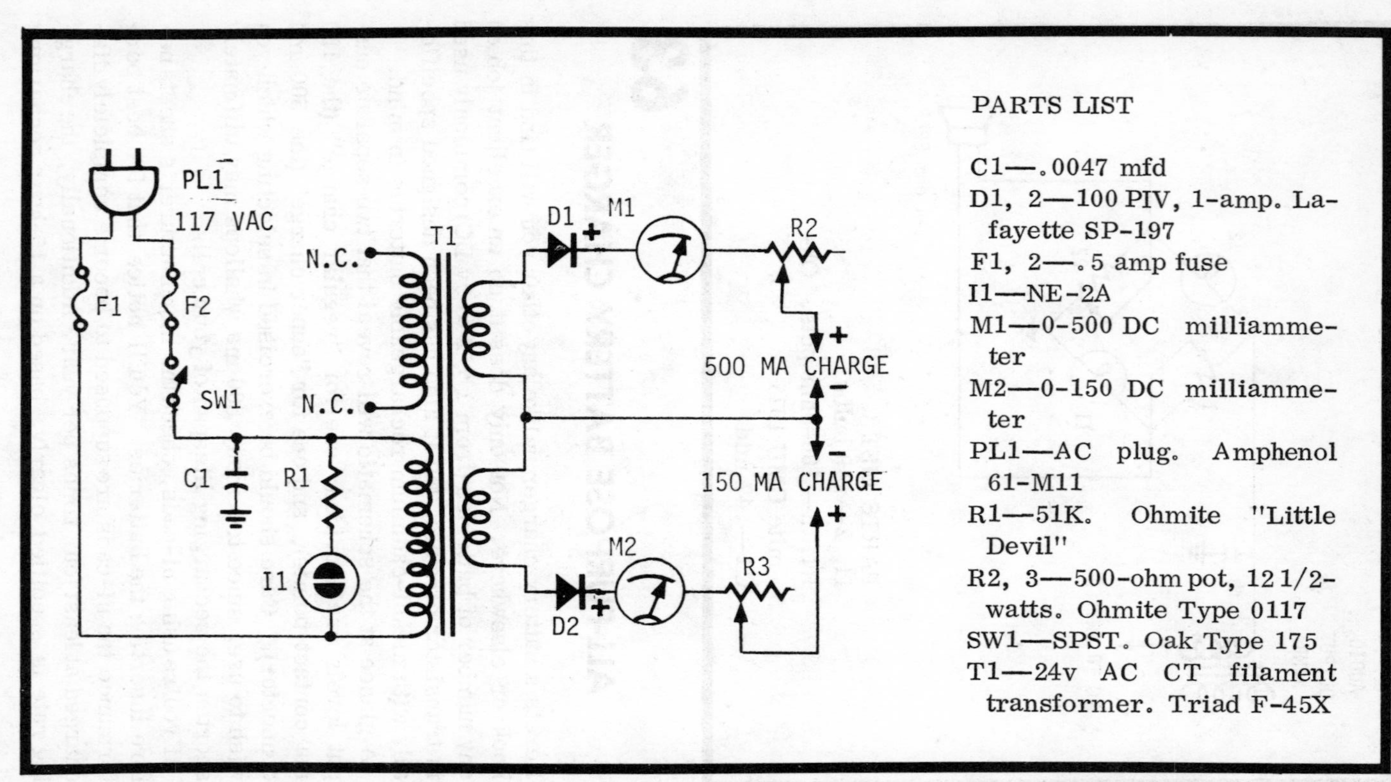

PARTS LIST

C1—.0047 mfd
D1, 2—100 PIV, 1-amp. Lafayette SP-197
F1, 2—.5 amp fuse
I1—NE-2A
M1—0-500 DC milliammeter
M2—0-150 DC milliammeter
PL1—AC plug. Amphenol 61-M11
R1—51K. Ohmite "Little Devil"
R2, 3—500-ohm pot, 12 1/2-watts. Ohmite Type 0117
SW1—SPST. Oak Type 175
T1—24v AC CT filament transformer. Triad F-45X

period. Never give a battery rated for a 150-ma charge a "super charge" unless you want to see a small explosion.

You can monitor how your battery charge is progressing by eyeing either M1 or M2 (or both). Yes, it is possible to run both sections at once with no sacrifice. When the needle appears to have leveled off, test your battery. It's probably ready to use again.

64

"BEEPER" TRANSISTOR TESTER

Here's a transistor tester for the real electronics experimenter who could care less about various specifications but just wants to know whether the transistor he has on hand is "good" or "bad" before he uses it to build a project.

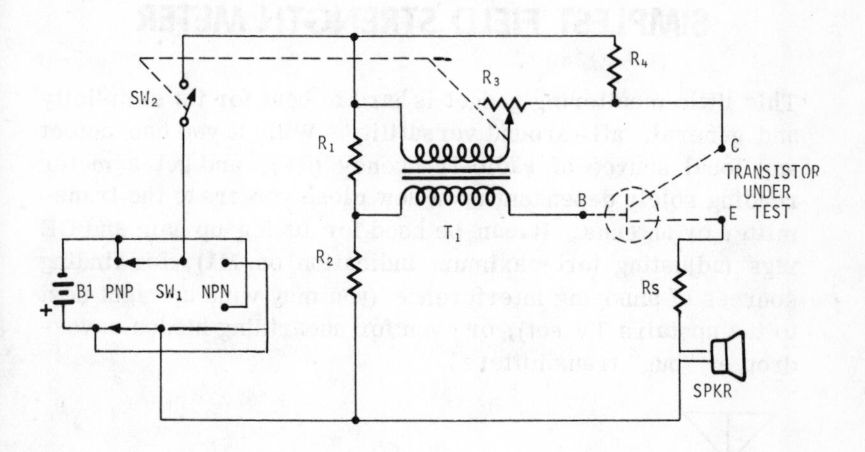

PARTS LIST

B1—3v DC

R1—4K. Ohmite "Little Devil"

R2, 4—1.1K. Ohmite "Little Devil"

R3—50K pot. Ohmite CMU 5041

R5—360. Ohmite "Little Devil"

SW1—DPDT. Oak Series 200

SW2—SPST on R3. Ohmite CS-1

T1—TR radio output transformer, 5K: 100 ohms. Argonne AR-111

Well, this little gadget does just that. If you hear a shrill "beep" in the speaker, your transistor is good. No sound at all and the transistor can be thrown on the scrap pile.

Construction is simple and straight-forward. The unit can be made to mount quite neatly in a 7 x 4 x 2" bakelite case with a metal lid that screws down.

Notice that switch SW1 allows you to alternate the circuit for both PNP and NPN transistor types. Always check first to see what type your questionable transistor is.

In operation, merely adjust R3 until a "beep" is heard. This is best accomplished by starting from a wide-open position and slowly "tuning" backwards towards zero-resistance.

65

SIMPLEST FIELD STRENGTH METER

This little monitoring gadget is hard to beat for its simplicity and general, all-around versatility. With it you can detect any local source of radio-frequency (RF), and get a meter reading solely dependent upon how close you are to the transmitter or antenna. It can be used for tuning up ham and CB rigs (adjusting for maximum indication on M1), for finding sources of annoying interference (you may wind up right next to the upstairs TV set), or even for unearthing hidden eavesdrop or "bug" transmitters!

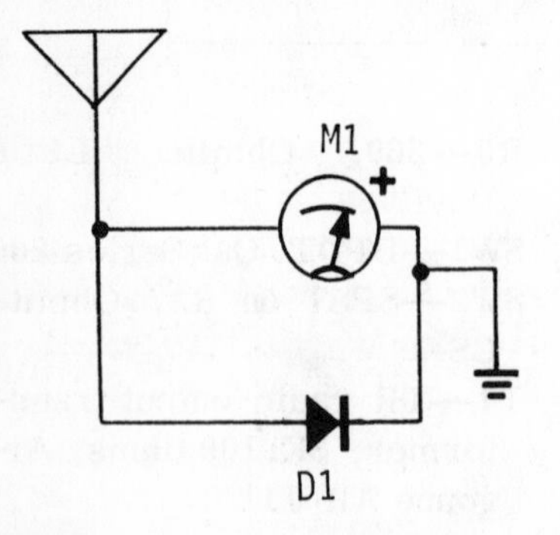

PARTS LIST

D1—1N38B
M1—0-1 DC milliammeter

The whole thing can be built into a case exactly the size of the meter. Subminiature meters often come with small triangular-shaped bakelite cases that are ideal for this purpose. Many experimenters ignore the "case" idea altogether and simply carry the device about by the meter. The antenna you'll want is dependent upon how far you are from the RF source. If you'll be doing close test work on ham transmitters, a single 10-turn coil of wire with one end connected to the meter will do the job. For unknown transmitter finding and antenna tune-ups, you might want to use a TR-radio type telescoping whip.

66

ELECTRONIC REFLEX-ACTION GAME

Perhaps this project could be called a "fast-on-the-draw" game, too, or any number of other possibilities. The principle is interesting; it's based on the famous 15¢ horse race games at Coney Island, and many traveling carnivals, where the contestants (usually a minimum of 4) are tested for their reaction to a horse-race taking place on a brightly-colored screen in front of them.

When a buzzer sounds, the player first hitting the switch sees his horse (and no other) advance one length toward the goal. Then music is pumped through the PA system and suddenly the buzzer signals another race to the switches, which are generally large levers or pushbuttons. Whoever gets to his switch first gets another length ahead in the race for a box of stale salt water taffy.

In this project there are no racing horses, but of all the lights presented on your board only one will light and that will belong to the player who hit his switch before all the others. After the first light, nothing happens regardless of how fast the other players may be. This circuit reacts to voltages switched in microseconds. And it can't be fooled. For excitement, offer a prize to the first player to get three consecutive wins. Then sit back and watch the action.

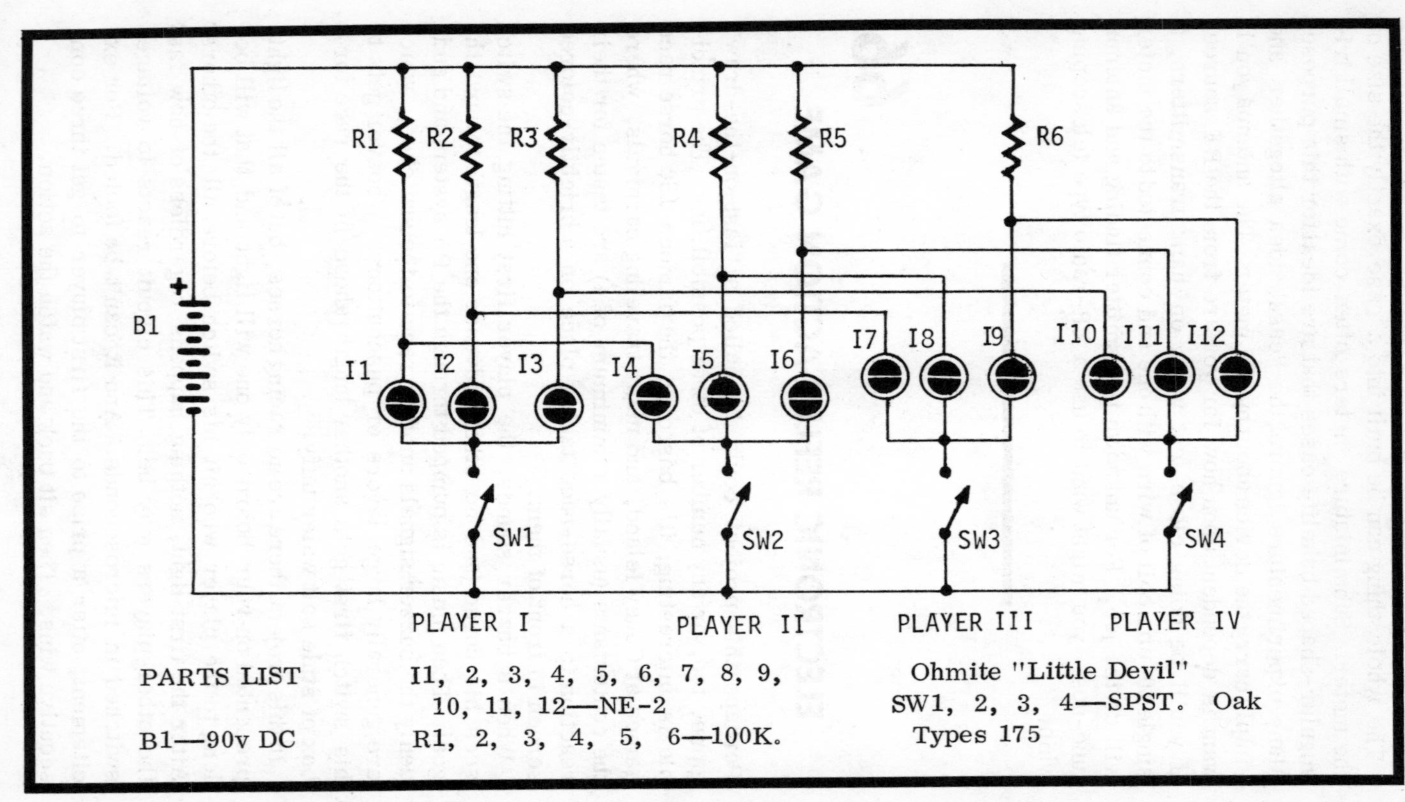
B1
+
R1
R2
R3
R4
R5
R6
I1
I2
I3
I4
I5
I6
I7
I8
I9
I10
I11
I12
SW1
SW2
SW3
SW4
PLAYER I
PLAYER II
PLAYER III
PLAYER IV
PARTS LIST
B1—90v DC
I1, 2, 3, 4, 5, 6, 7, 8, 9,
10, 11, 12—NE-2
R1, 2, 3, 4, 5, 6—100K.
Ohmite "Little Devil"
SW1, 2, 3, 4—SPST. Oak
Types 175

67

KILOWATT RF VOLTMETER

Here's a device guaranteed to warm the heart of any ham radio operator: An in-line RF meter that will handle the full legal power level—1000 watts—as well as lower-powered transmitters in the 20-watt-and-under classification. Also, the meter permits your seeing exactly what's going up to the antenna as opposed to an FSM which may be reading stray RF or a harmonic instead of the actual output signal.

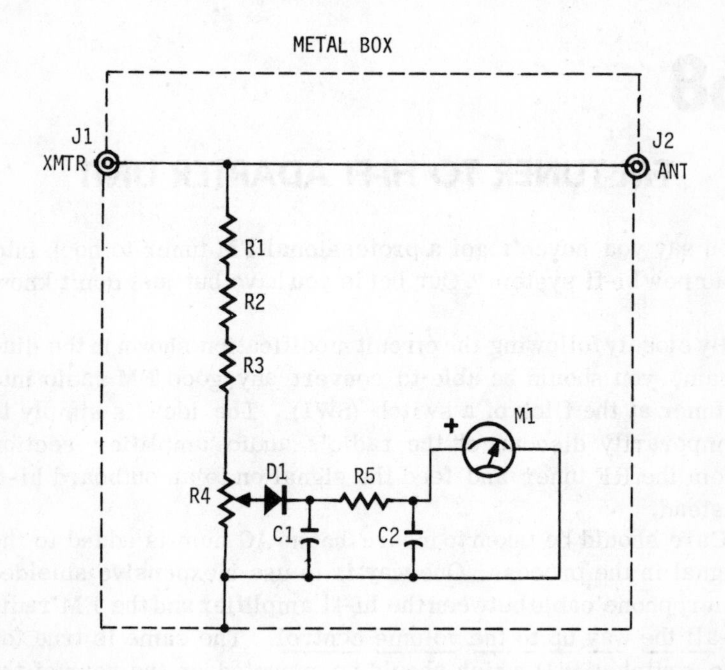

PARTS LIST

C1, 2—.0047 mfd
D1—1N38B
J1, 2—SO-239 coax receptacles. Amphenol 83-1RTY
M1—0-1 DC milliammeter
R1, 2, 3—2.4K, 2 watts. Ohmite "Little Devil"
R4—5K pot, 2 watts
R5—10.1K, 2 watts Ohmite "Little Devil"

Built into a small universal-type meter box (which can measure 4 3/4 x 4 1/4 x 4 1/4"), the voltmeter is outfitted with two standard SO-239 coax connectors. The device can be used with any matched 75-ohm or 50-ohm antenna system.

In use, merely adjust R4 for a comfortable setting and tune the transmitter's output loading controls for a maximum reading on M1. If you use a long-wire antenna, ground the box to your transmitter with a section of heavy copper wire.

Notice that R4 must be a composition (conventional) type potentiometer—not one of the wirewound variety.

68

FM TUNER TO HI-FI ADAPTER UNIT

You say you haven't got a professional FM tuner to hook into your new hi-fi system? Our bet is you have but just don't know it.

By closely following the circuit modification shown in the diagram, you should be able to convert any good FM radio into a tuner at the flick of a switch (SW1). The idea is simply to temporarily disconnect the radio's audio amplifier section from the RF tuner and feed the signal on to an outboard hi-fi instead.

Care should be taken to insure that no AC hum is added to the signal in the process. One way is to use inexpensive shielded microphone cable between the hi-fi amplifier and the FM radio —all the way up to the volume control. The same is true for the switch (SW1) which should be mounted on the rear of the FM radio and not at the amplifier for best results.

PARTS LIST

SW1—SPST. Oak Type 175

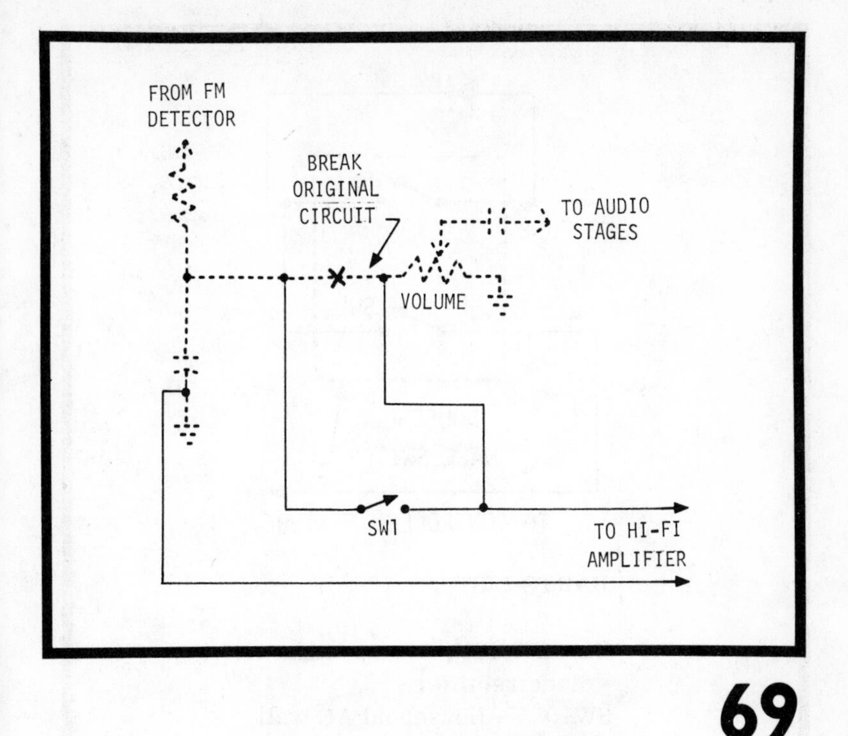

69

$1.50 LIGHT DIMMER

Although most hardware stores, electronic parts outlets, and electrical supply stores will sell you a commercial version of this project, chances are it will cost quite a bit. Diodes rated at 3 amps with PIVs of 200 and higher are being distributed through many surplus houses for around 88¢ these days, and even brand-new from a major electronics company they shouldn't run over $1.50. Chances are, however, that you'll have this one somewhere in your junkbox just waiting to be pressed into action.

To make this effective light dimmer work, all you need is a replacement-type double wall switch upon which you can mount D1 as indicated in the accompanying diagram. Once completed, you have converted the switch into a lamp dimmer that looks no different than a conventional wall switch.

In operation SW2 (lower switch) will turn the lights on and off; SW1 will dim them to one-half normal brilliance when it is open and provide maximum light when closed. The 3-amp rectifier can control a maximum total wattage of about 250.

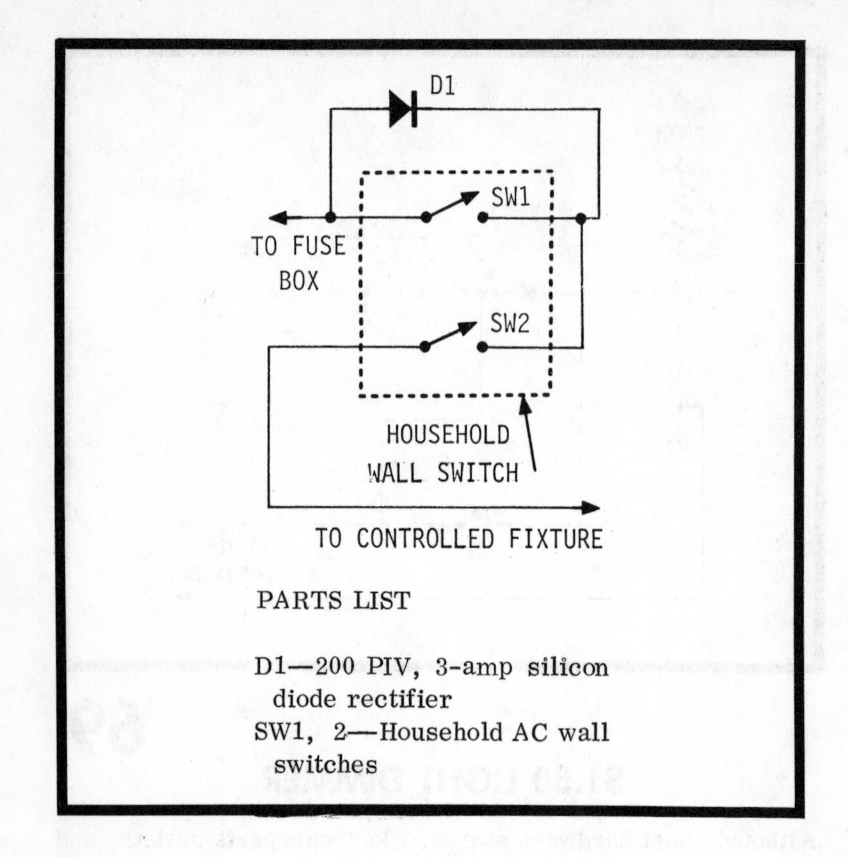

70

SCR DIODE CHECKER

Here's another junkbox great: a fail-safe silicon rectifier diode tester that performs the one function important to any true electrons experimenter—telling you if the SCR is good or bad. No nonsense, no computing, no complicated graph interpretations. Just plug it in and find out what you want to know.

For expediency's sake, we've used Westinghouse 1N1217 diodes for D1 and D2. You may want to substitute, but before you do we suggest your checking with some of the companies listed in the back of the book. All other components are fairly conventional types, easily scrounged from a number of sources.

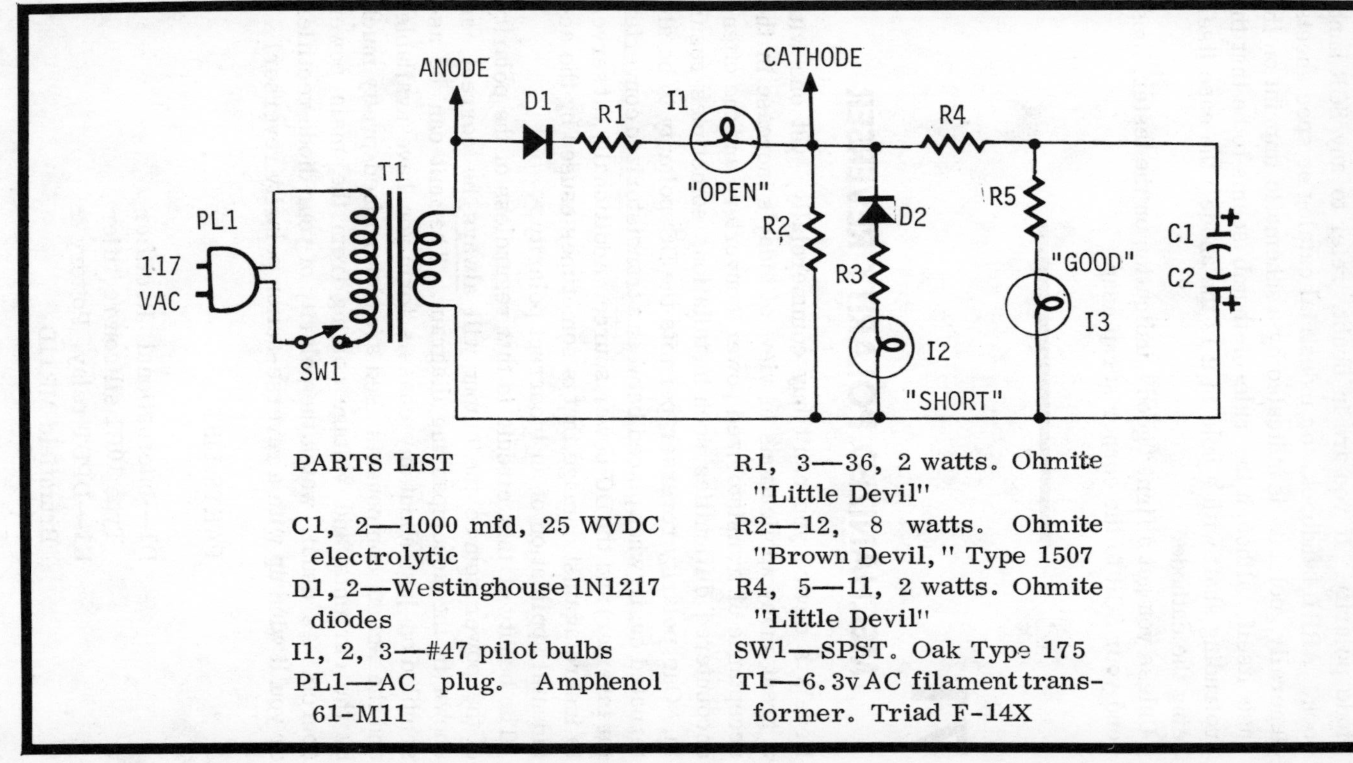

PARTS LIST

C1, 2—1000 mfd, 25 WVDC electrolytic

D1, 2—Westinghouse 1N1217 diodes

I1, 2, 3—#47 pilot bulbs

PL1—AC plug. Amphenol 61-M11

R1, 3—36, 2 watts. Ohmite "Little Devil"

R2—12, 8 watts. Ohmite "Brown Devil," Type 1507

R4, 5—11, 2 watts. Ohmite "Little Devil"

SW1—SPST. Oak Type 175

T1—6.3v AC filament transformer. Triad F-14X

Before throwing switch SW1, make absolutely sure of proper diode polarity. If you are in doubt, refer to any SCR handbook, ARRL handbook, or industrial catalog or spec sheets. Generally, polarity is indicated by a schematic imprint on the case itself, although the rule-of-thumb seems to be that the protruding shaft with a hole in it is the anode, the case itself being the cathode.

Unless you get a firm "good" indication on the tester, best send your SCR to the corner scrap heap.

71

MECHANIZED POLARITY REVERSER

Here's a device you can't buy commercially, yet one with a great many applications in view of today's increase in the acceptance of transistorized power converters and the often-encountered difficulties such installations sometimes result in. Our polarity reverser permits the DC polarity to be alternated easily when placed between a transistorized converter (or inverter) and the DC power source. Additionally, it serves to insure against damage that is sometimes caused by the accidental application of an incorrect polarity.

The beauty of this circuit is that regardless of the polarity of the power source the output will always be correct, as shown in the accompanying diagram. The unit can be assembled on just about any chassis form you have available, and the actual components used are determined pretty much by the currents and voltage coming from the main power source. As usual, watch the polarity of your diode rectifier or you'll wind up with a reverse-action polarity reverser!

PARTS LIST

D1—International Rectifier Type 10B1 silicon rectifier
K1—DPDT relay. Pottery & Brumfield MR11D

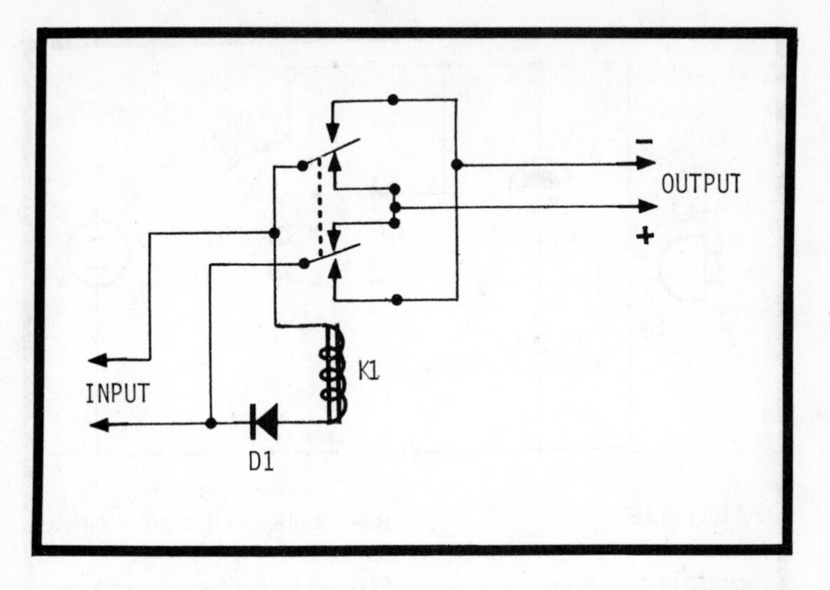

72

"INSTANT-ON" APPLIANCE ADAPTER

Being accustomed to all kinds of "instant-on" appliances, it's somewhat disconcerting to have to wait 20 seconds or longer for your old tube-type radio to warm up. Hence the current rash of commercial items selling anywhere from \$2.95 to \$8.95 as "instant-on" converters. Such devices are fine if you want to spend the money, but we prefer our own homebrew concoction that does the same job entirely with junkbox parts.

You're probably wondering how this thing works, but since this is not a textbook you'll have to dope it out yourself, if you're so inclined. Suffice to say that this adapter's been designed for garden-variety AC/DC kitchen radios and that the NE-2A is strictly an indicator to let you know that the radio's warmed up. (It's not functional in any way in the instant-on circuit.)

Incidentally, you can have a lot of fun by building this circuit <u>into</u> the AC/DC (omitting the neon assembly), disguising D1 as an electrolytic. The result (if you don't mind having the set warmed up all the time) is guaranteed to baffle your electronic-minded friends—particularly the so-called "experts."

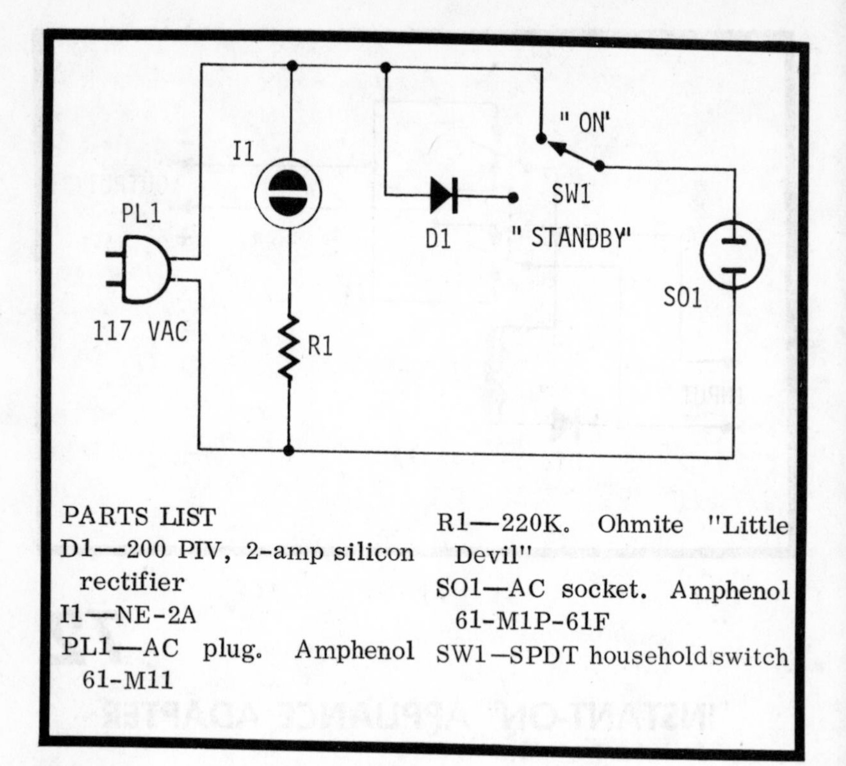

PARTS LIST
D1—200 PIV, 2-amp silicon rectifier
I1—NE-2A
PL1—AC plug. Amphenol 61-M11
R1—220K. Ohmite "Little Devil"
SO1—AC socket. Amphenol 61-M1P-61F
SW1—SPDT household switch

73

CHEATER CORD CURE-ALL

If you've ever done any TV servicing, you know what a nuisance those interlocks are when it comes to removing the back of the set. What usually happens is that you have to remove the back, disconnect the AC line from the wall socket, hunt about for a cheater cord and try to find an outlet close enough to hook up the cord (which is usually too short) and your test instruments and soldering gun.

This simple project, however, will put an end to such frustration and fuming. Complete with a neon-assembly pilot light that tells you when the Cure-all is on, the control box contains its own AC socket. Hence you've just eliminated 20 minutes of scrambling every time there's a servicing job, plus a multitude of extension cords.

The Cure-All can be designed to fit into just about any size container you have handy, since the biggest item physically is the accessory socket. If you like, you can string two or three accessory sockets into the circuit, thereby doubling and tripling the versatility of your finished unit.

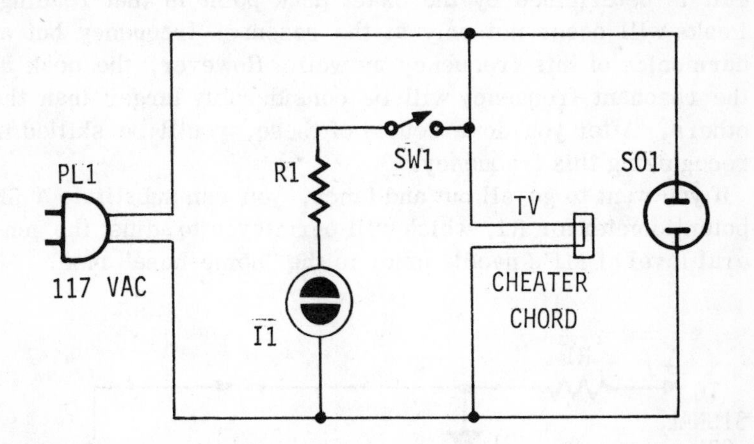

PARTS LIST

I1—NE-2A
R1—51K. Ohmite "Little Devil"
PL1—AC plug. Amphenol 61-M11
SO1—AC socket. Amphenol 61-M1P-61F
SW1—SPST. Oak Type 175

74

AUTOMATIC FREQUENCY FINDER

Here's a circuit that's been making the rounds in certain VHF ham circles for years, yet has an application for just about any tinkerer who likes to roll his own. What it does is determine the resonant frequency of just about any parallel L/C (coil-capacitor paralleled) circuit you have going. All it requires is a 0-50 DC microammeter, a single 2600-ohm resistor, a .0022-mfd capacitor, and a run-of-the-mill-type diode detector (1N38B). Connect the Frequency Finder to any available RF signal generator and you're in business.

In operation the signal generator is tuned very slowly through its entire range of frequencies as you watch for a sudden peak reading on the meter (M1). Exact frequency of the tuned circuit is determined by the exact peak point in that reading. Peaks will occur not only at the resonant frequency but at harmonics of this frequency as well. However, the peak at the resonant frequency will be considerably larger than the others. After you do a couple of these, you'll be skilled in recognizing this frequency.

If you want to go all out and fancy, you can substitute a 5K potentiometer for R1, which will permit you to adjust the general level of M1's needle prior to the "home-base" rise.

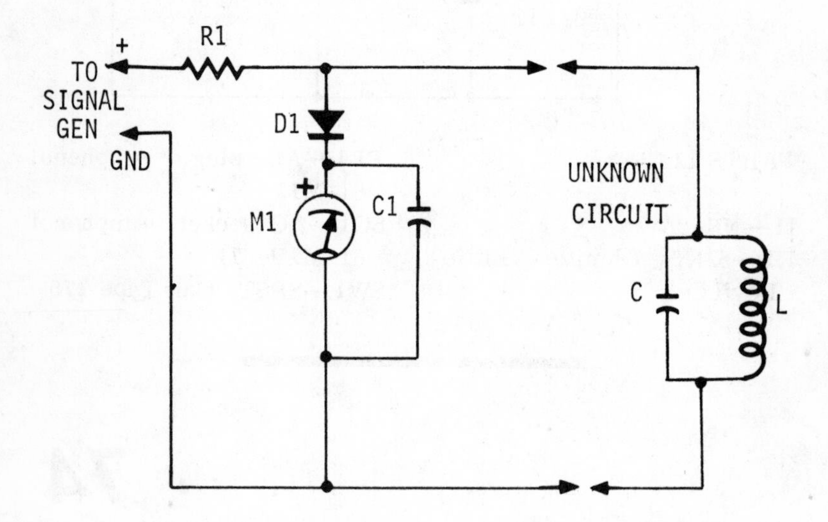

PARTS LIST

C1—.0022 mfd
D1—1N38B
M1—0-50 DC microammeter
R1—2.7K. Ohmite "Little Devil"

75

WORLD'S CHEAPEST INTERCOM

Though you may have not been aware of the fact, acceptable intercom performance canbe achieved with only apair of PM loudspeakers connected together. That's right! No transistors, ICs, tubes, or other components required!

Trick to optimum performance, however, is in selecting PM speakers with the largest magnet available (typically, these range from 0.66 to 1.5 ounces or more). The heavier the speaker magnet, the more volume you can expect from the intercom. Of course, the volume is not to be compared with intercoms containing several stages of amplification but it is sufficient, generally, for quiet locations, such as in most households. It is not recommended, obviously, for noisy factory areas.

To boost output, you can install universal 500-ohm primary and 3.2-4-ohm secondary line-to-voice coil transformers at the speakers (primary to speakers, secondary to interconnection wires). But first, try your intercom without the transformers. You may be surprised at the performance!

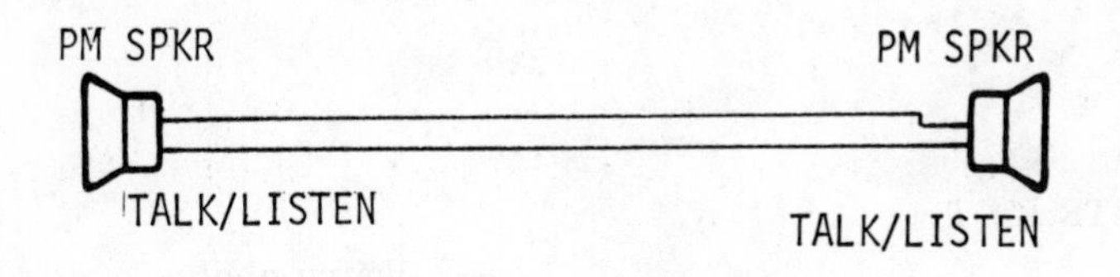

PARTS LIST

2—Four - inch diameter PM loudspeakers with large Alnico-V magnets. See text

76

150-VOLT UNIVERSAL POWER SUPPLY

Here's a dandy power supply which can be whipped together in no time almost completely from junkbox materials. It can be designed to fit any available chassis and will save you hours of exasperation when facing a construction project involving a 1- or 2-tube circuit requiring 150v DC B-plus voltage. Initially constructed to power a VHF converter, it wound up as a receiver supply for a 2-tube 80-meter project.

The filter capacitors, which are electrolytic, can be placed individually, or as a multi-section can. Values are not critical so long as you watch WVDC ratings.

If you like, you can add a 3-watt 65K bleeder resistor across the B-plus output and ground. This will be helpful if you're not feeding into a constant-load.

Additionally, you can rig up a pilot lamp across the filament winding or even a neon bulb assembly across the primary of T1. The author suggests bringing the output to a three-screw terminal strip for maximum versatility.

PARTS LIST

C1—40 mfd, 350 WVDC electrolytic
C2—20 mfd, 350 WVDC electrolytic
C3—25 mfd, 350 WVDC electrolytic
D1—325 PIV, 60-ma silicon
L1—8-henry choke
PL1—AC plug. Amphenol 61-M11
R1—51, 2 watts. Ohmite "Little Devil"
SW1—SPST. Oak Type 175
T1—150 & 6.3v AC. Merit P-3046

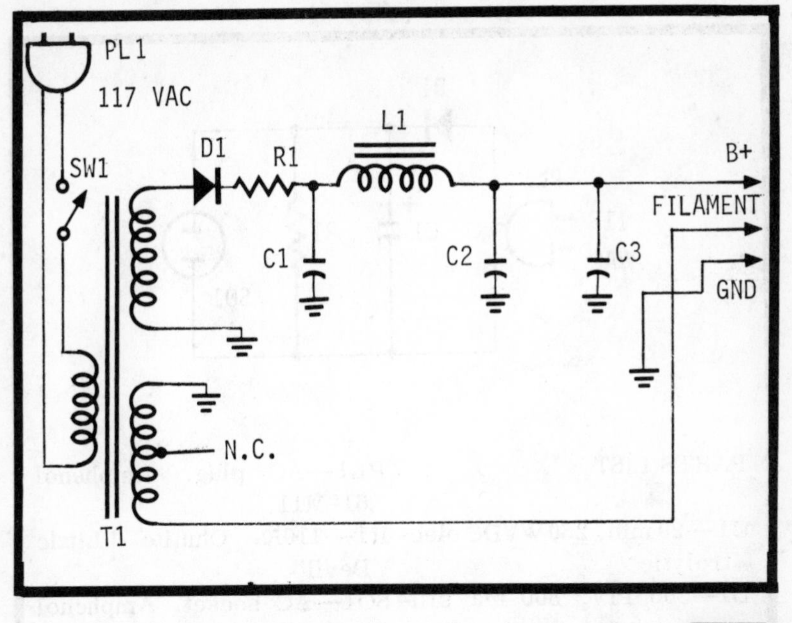

77

AN ELECTRONIC SHAVER CONVERTER

Many people believe that the best way to get a really good shave with an electric razor is to power the device with DC instead of the conventional household 117v AC. The reason many give is that the shaver runs more efficiently and smoothly, which is said to result in a greatly-improved morning shave.

This device, while extremely simple to build, has been designed to supply your electric razor with half-wave DC. If you doubt the concept, we dare you to build it and give the Shaver Converter a comparison check with normal AC use. Our bet is you'll never go back to old-fashioned (converterless) shaving.

Bear in mind that your converter will be used in the bathroom where it is very likely to get wet, if not extremely moist. Seal the converter thoroughly and place it where the kiddies won't find it (if that is possible!).

Final note: Mark clearly on your chassis "For Shaver Only." Connecting other appliances to a rectified DC source could prove disastrous.

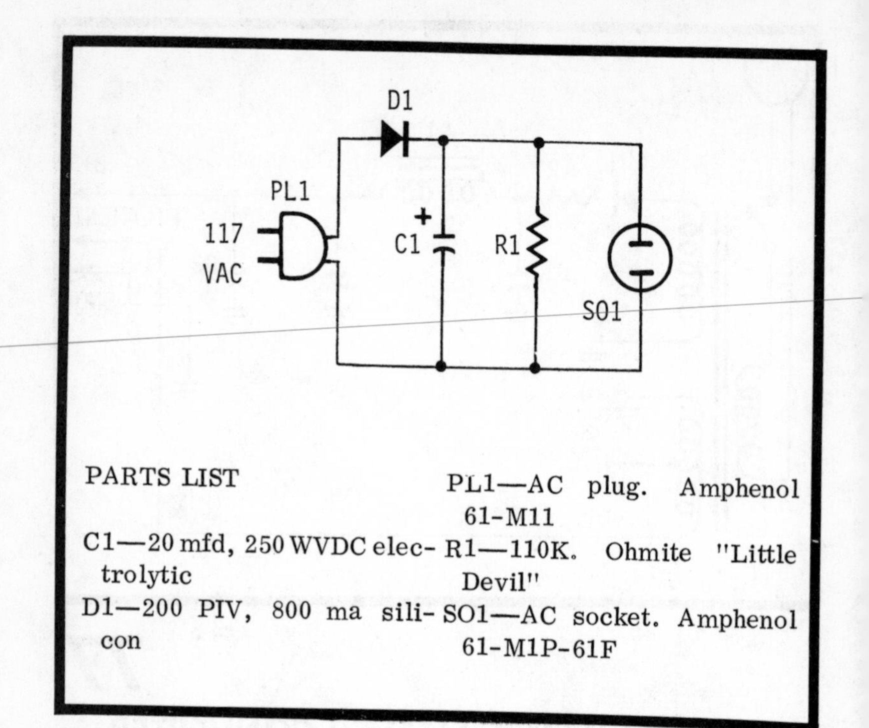

PARTS LIST

C1—20 mfd, 250 WVDC electrolytic

D1—200 PIV, 800 ma silicon

PL1—AC plug. Amphenol 61-M11

R1—110K. Ohmite "Little Devil"

SO1—AC socket. Amphenol 61-M1P-61F

78

SHAVE-IN-YOUR-CAR ADAPTER

If you don't plan on plunking down $29.95 for a commercial transistorized power inverter for your automobile, yet still like the idea of being able to shave on your way to work in the morning, this gadget is for you. Operating on the same general principle as the converter in the last project, this inexpensive adapter makes use of your car radio's B-plus circuit to power the shaver. Of course, if you're using a new transistorized radio that runs completely off 12v DC, forget it. But check the schematic of the radio (pasted inside the removal back plate) and see. A great number of these radios still use power ranging to 150v DC, achieved through transformers and solid-state vibrator replacements. Regardless, see if you can tap off at least 117v AC.

Should you happen to have a tube radio or CB set from which you can steal 250 volts B-plus, you can use this, too. Simply adjust R1 for a voltmeter reading that does not exceed 120v with the accelerator floored. Even if the most you can steal is 80-90 volts, don't despair. You'd be surprised what a shave you'll get anyway.

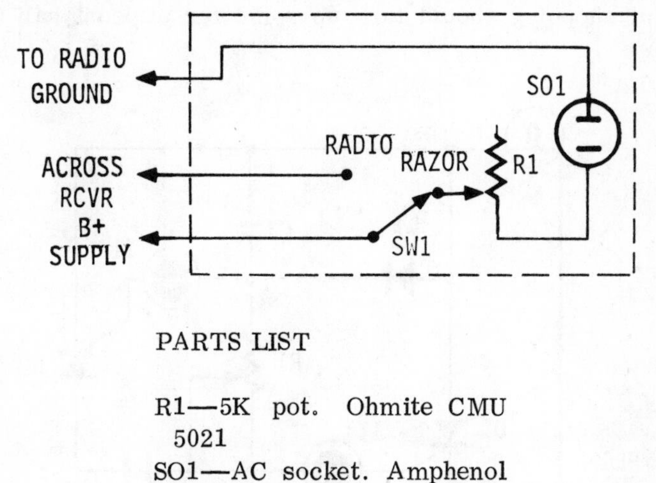

PARTS LIST

R1—5K pot. Ohmite CMU 5021
SO1—AC socket. Amphenol 61-M1P-61F
SW1—SPDT. Oak Series 200

79

ROMANTIC LIGHT DIMMER

Undoubtedly you've seen the rheostat-type wall light switches which permit your dimming the overhead lights to any desired level at the flick of the dial. Unfortunately, however, these things seem to retail at prices that generally <u>start</u> at $7.95 and go all the way to $16.95 for some outlets. Naturally, not desiring to go this route, we checked out some of the whole-sale components people (see this list in the back of the book) and found that by careful buying we could roll our own for considerably less. With junkbox parts and wholesale-priced

SCRs it shouldn't run you more than $3.00 tops. A further saving is realized by employing NE-83s (see the schematic).

For dimming fluorescent lamps, always "start" the tube by beginning at maximum brilliance, then slowly dimming it until the desired level is attained. CAUTION: Never use on high-wattage loads.

Incidentally, list price of the RCA 2N3528s is $1.63. But careful shopping should turn up some for at least half this price.

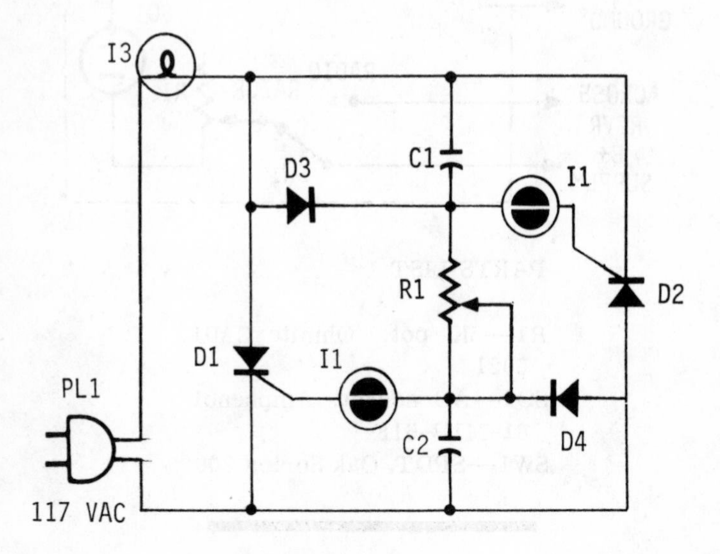

PARTS LIST

C1, 2—.22 mfd
D1, 2—2N3528 SCR. 1.3 amp, 200 PIV
D3, 4—1N4002
I1, 2—NE-83
I3—Incandescent or floures-cent bulb to be regulated, up to 200-watt rating
PL1—AC plug. Amphenol 61-M11
R1—100K pot. Ohmite CMU 1041

80

TAPED CQ'ER DEVICE

Here's a contester's delight: an adapter which will allow you to call "CQ Contest" to your heart's content automatically by means of an endless loop on your tape recorder. Nothing new? Perhaps not at first glance (VHF phone men have been doing it for years), but a close inspection reveals that we're talking about a device for CW!

The adapter, which can be built into your transmitter or as a separate unit altogether, converts modulated pulses into a relay action that in turn keys the transmitter. To operate, you merely program a "CQ CONTEST DE K2ZSQ" type endless tape (a continuous loop) of CW stamped out on a good code practice oscillator. The output (MCW) is then coupled directly to the adapter, where it actuates the relay.

After the system has been hooked up, all that remains is for you to adjust the volume control of the recorder for proper relay response. It should be noted, of course, that a key will still be necessary for you to interject your long "K" sign-off, although this can be eliminated if you elect not to use a continuous tape but instead a single "CQ DX" tape that runs to the very end of the reel. This can usually be accomplished by making the recording, playing it back, and snipping the programmed section of the tape (perhaps 5 to 7" worth). This should then be inserted as a separate tape into the recorder with the end of the tape firmly affixed to the spool by means of Scotch Tape. In this way you won't lose your tape when it runs out. If you have an automatic cut-off feature on the recorder, you're in business. If not, you'll have to shut off the mechanism manually. This isn't any real problem, however, because by this time you're getting replies to your CQ and the other chap could care less about the recorder. It's done its job!

A slightly more sophisticated approach that is particularly applicable at the VHFs is to program an entire tape with "CQ CONTEST DE K2ZSQ (three times) K" followed by a 30-second pause and then beginning again. The pause, which you can vary to suit your own tuning habits, serves as the time during which you will listen for replies. A VOX circuit is directly hooked to the recorder output (paralleled with the keying adapter) which keeps the recorder running only when it is sending your transmission. During the pause periods, it shuts off your transmitter completely and switches you to receive. If you don't get a reply, it automatically starts the transmitter again and kills the receiver B-plus (standby switch). If you do, you simply stop the recorder.

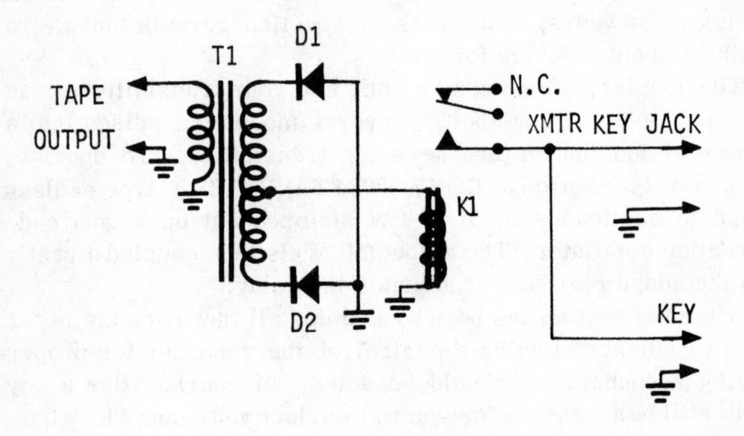

D1, 2—125 PIV, 50-ma silicon
K1—2.5K SPDT relay. Potter & Brumfield GB5D
T1—Universal output transformer. Lafayette 33G7505

81

"THE VACATIONER" AUTOMATIC HOUSE SWITCH

Want a foolproof method for foiling would-be thieves and vandals? This inexpensive gadget makes use of a photoelectric cell to "sense" darkness and to turn on small lamps throughout the house come evening, thereby automatically giving the impression that the occupants are indeed home and active.

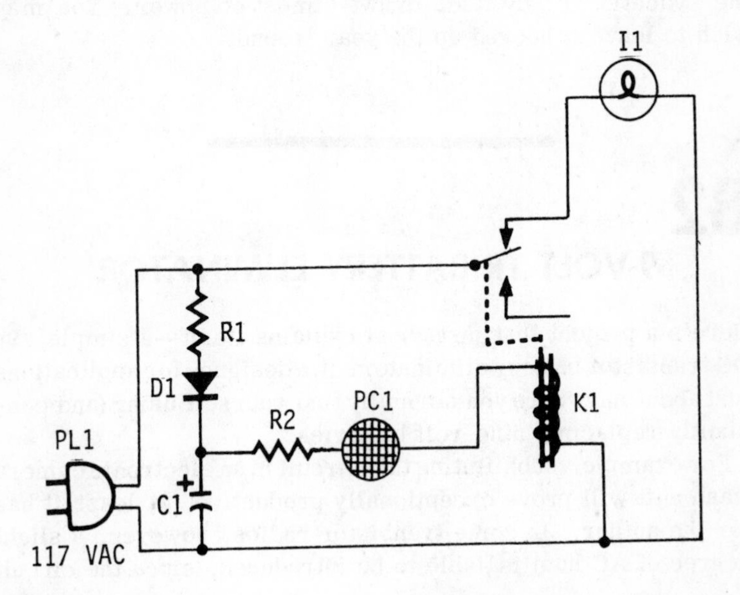

PARTS LIST

C1—4 mfd, 350 WVDC electrolytic

D1—RCA 40265. 130 ma, 400 PIV rectifier

I1—Household lamp or lamps, to 100 watts total

K1—24v DC SPDT relay. Newark 60F1749

PC1—Photocell

R1—16

R2—11K, 3 watts

Actually, you'll notice that the circuit says something different. What really happens is that the lamps are always on, except when sunlight hits the photocell. The inexpensive relay functions as the trigger which in turn breaks contact to the electrical circuit.

Of course, you could adapt this automatic switch arrangement for some other purpose, or in fact use it to power any 117v AC device which doesn't draw more than 100 watts.

Be sure that the photocell is placed (yes, you can remote it over a considerable distance) in such a position that it always gets sunlight. You wouldn't want the lights to come on at 2:30 in the afternoon when the photocell "moved" into a shady spot!

Construction is simple, yet remember that you're dealing with house current—not 9-volt TR batteries. Make certain that R2 is at least 3 watts and nothing smaller. Incidentally, the "vacationer" by itself draws almost no power. You may wish to leave it hooked up the year 'round.

82

9-VOLT TR BATTERY ELIMINATOR

Here's a project that just about explains itself—a simple, 9v DC transistor battery eliminator. It's designed for applications just about anywhere you normally find yourself using (and constantly replacing) nine-volt batteries.

For example, substituting this circuit in an electronic camera flash unit will prove exceptionally productive, at least it has for the author. In some transistor radios, however, a slight degree of AC hum is liable to be introduced, since the circuit

PARTS LIST

C1—30 mfd, 150 WVDC electrolytic

D1—30- or 50-ma selenium rectifier

PL1—AC plug. Amphenol 61-M11

R1—51, 1 watt. Ohmite "Little Devil"

R2, 3—6.8K, 2 watts. Ohmite "Little Devil"

shown in the accompanying diagram employs a half- rather than a full-wave rectified supply. In other radios we experimented with, however, no traceable hum was detected.

In any case, exercise caution; you will be dealing with 117v AC and will want to keep all leads insulated. House the whole unit in such a way as to preclude accidental shock.

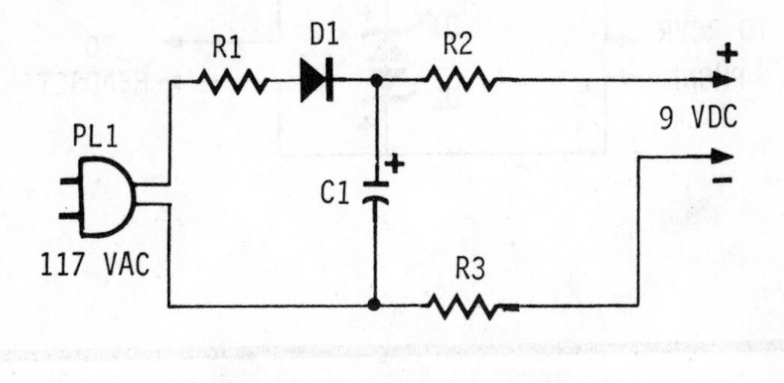

83

SIMPLEST NOISE LIMITER

Here's an amazingly effective noise limiter you can put together in about 15 minutes, yet it will save you plenty of frustration if you're an avid DXer or SWL.

Using two inexpensive zener diodes connected more or less back to back, the limiter will eliminate some of those ear-splitting occasions when a strong local signal suddenly plops in on frequency right on top of your S2 DX station. Additionally, it'll allow you to tune to your own frequency when transmitting CW (to monitor your signals and keying) without necessitating your dashing across the table—spilling the 807 all over the logbook—to turn the volume nearly all the way down.

Merely connect the gadget between the receiver and headphones and you're in business. Chassis? Any plastic box or pill container will do.

PARTS LIST

D1, 2—6.8-volt, 400-milliwatt zener. Motorola 1N957

R1—11K. Ohmite "Little Devil"

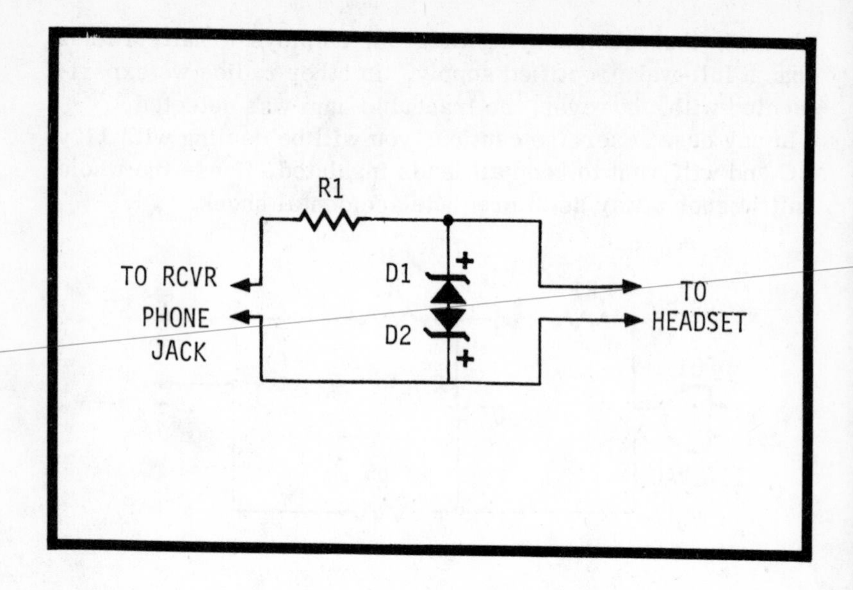

84

SLIDE-SWITCH LOUDNESS CONTROL

When you've been suddenly called away from the record player and reduce the volume to take a telephone call, did you ever notice how the bass and treble response no longer seem as good? Seldom is this caused by any defect in your player's electronics; rather, it is inherent in human hearing that certain volume levels (higher, generally, than low background music) are necessary before the ear "hears" the presence of bass and certain treble notes.

The circuit shown in the accompanying diagram, however, takes this into consideration. Since bass notes are most often adversely affected by a sudden loss in volume, they are boosted the most. What happens, then, is that as SW1 is switched on the loudness control takes over, simultaneously lowering volume from concert hall levels to soft, background presence while retaining the bass (and treble) proportions to which you were accustomed at the higher volume setting.

In construction, keep leads short, house the unit in a shielded chassis, and use shielded mike cable to interconnect between your preamplifier and final amp.

If a single-chassis preamp-amplifier is the case, disconnect the lead presently wired to the arm (center terminal) of the volume control and connect this wire to the input of the loudness control. The output, of course, goes to the volume control.

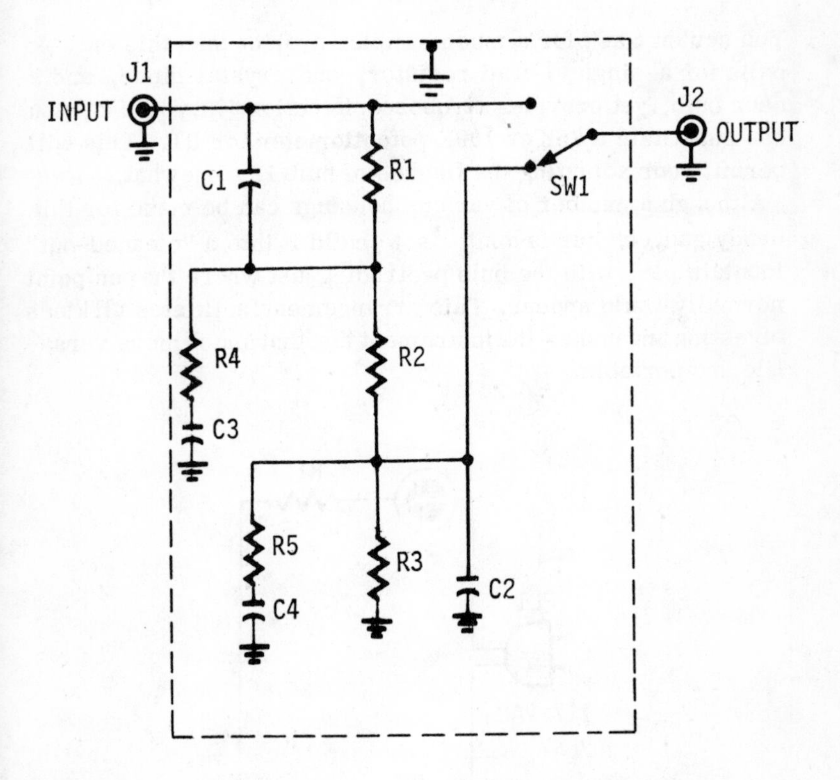

PARTS LIST

C1—.0015 mfd
C2—500 pfd
C3, 4—.015 mfd
J1, 2 — Audio connectors (receptacles), Amphenol 75-3
R1, 2, 3 — 33K. Ohmite "Little Devil"
R4, 5—Ohmite "Little Devil"
SW1—SPDT. Oak Type 200

85

99¢ STROBOSCOPE

You couldn't ask for a much simpler project than this one. It calls for a single 1-watt resistor, one crystal diode, and a neon bulb, yet provides stroboscopic action. If you like, you can substitute a 75K or 100K potentiometer for R1. This will permit your adjusting the flashes of bulb I1 somewhat.

Although a number of various housings can be made for this dandy gadget, our favorite is to build it into a "cleaned-out" fountain pen, with the bulb protruding just where the pen point normally would appear. This arrangement facilitates all kinds of testing and makes the instrument just that much more versatile and portable.

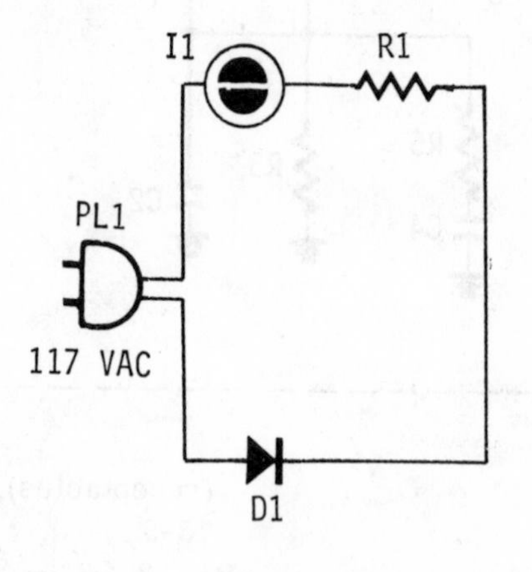

PARTS LIST

D1—1N38B
I1—NE-48
PL1—AC plug. Amphenol 61-M11
R1—36K, 2 watts. Ohmite "Little Devil"

86

12¢ LIGHTNING DETECTOR

Ridiculous as this may sound, some experienced amateur radio operators are not even aware of this simple trick. The key to best results with this 12¢ detector is a good, wet ground and a long, high antenna wire. The longer the antenna, the further away you'll be able to detect lightning.

If you find it necessary (as K2ZSQ once did, hooking his neon bulb to a 1200-foot long-wire antenna), you may want to add a 25K potentiometer across the antenna and ground, permitting you to adjust the brightness and flashes of the bulb under overload conditions. If you're very close to the storm (or simply have a darned good antenna), you'll find that the neon flashes seem to meld together into one long continuous glow. In this case, add some resistance by adjusting the pot.

PARTS LIST

I1—NE-2

87

NEON BULB AC-DC VOLTMETER

Here's a test instrument that at first glance seems similar to Project 84, until you realize that this gadget requires no power supply of its own. Instead, it merely samples both AC and DC voltages to tell you what kind of currents you're fooling around with.

By building it with a faceplate similar to that in Project 84 (pointer knob and all), you can construct it in a small plastic parts box and wind up with a truly pocket-size, portable testing device.

Calibration is identical to that described in the previous tester, except that you can make it simpler if you wish just by feeding in a known voltage and marking this on the faceplate. If you have a variable-voltage power supply handy, you're in business. If not, you'd do best with the technique shown in Project 84.

In use, adjust R1 until the neon light just lights, then read out the voltage on the dialplate. Incidentally, you can identify voltage polarity by keeping an eye on which electrode inside the neon bulb lights. Clue: The one that glows is invariably negative.

PARTS LIST

C1—.22 mfd
I1—NE-2
R1—5 meg pot. Ohmite CMU 5052
R2, 3—110K. Ohmite "Little Devil"

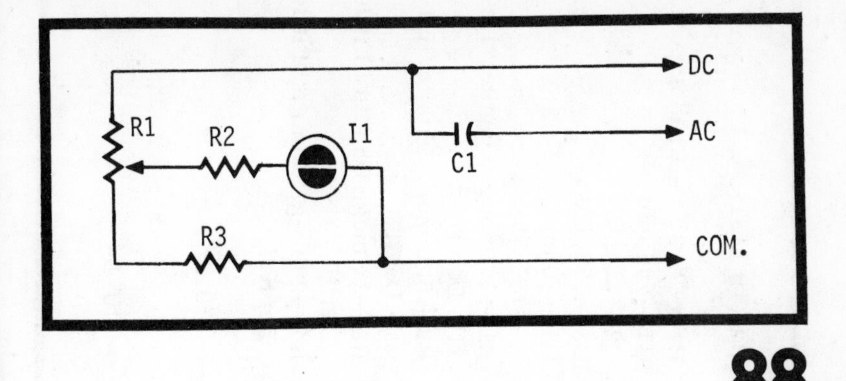

88

SIMPLE RC TESTER

No, this gadget won't tell you a thing about remote control airplanes, but if you need a gadget to help find values for junkbox resistors and capacitors, this baby's for you! It is another pocket-size test instrument that requires no elaborate meters or what-have-you, yet will read out satisfactorily a wide range of capacitance and resistance values to help you in locating that urgently needed component.

What'll it do? When accurately calibrated this simple tester will read resistances between 2.5 and 20 megohms and capacitances between .022 mfd and 1.0 mfd. Following earlier described techniques, make two separate face (or dial) plates for SW1 and SW2, marking the SW1 plate with ".05," ".1," ".25," and ".5," (mfd) and the plate for SW2 as "2.7," "5.1," "11," and "22" (megohms) to correspond to appropriate switch positions.

After completion, hook the RC tester to a power supply furnishing about 125v DC. This isn't critical; you can go down to about 95v DC and up to about 150v DC with no difficulty if you like.

For resistance finding, hook the unknown resistor across BP3 and BP4. Switch SW1 (on the capacitor side) to ".05 mfd." Now turn the power supply on, switch SW2 to the "R" position and watch the blinking of I1. Now rotate SW2 until you get the same pulse rate as you did with the unknown resistor across BP3 and BP4 (on position "R"), and you've got the value of the resistor. Simple? If the pulse rate is too high, try another setting with SW1. Incidentally, use common sense if all else

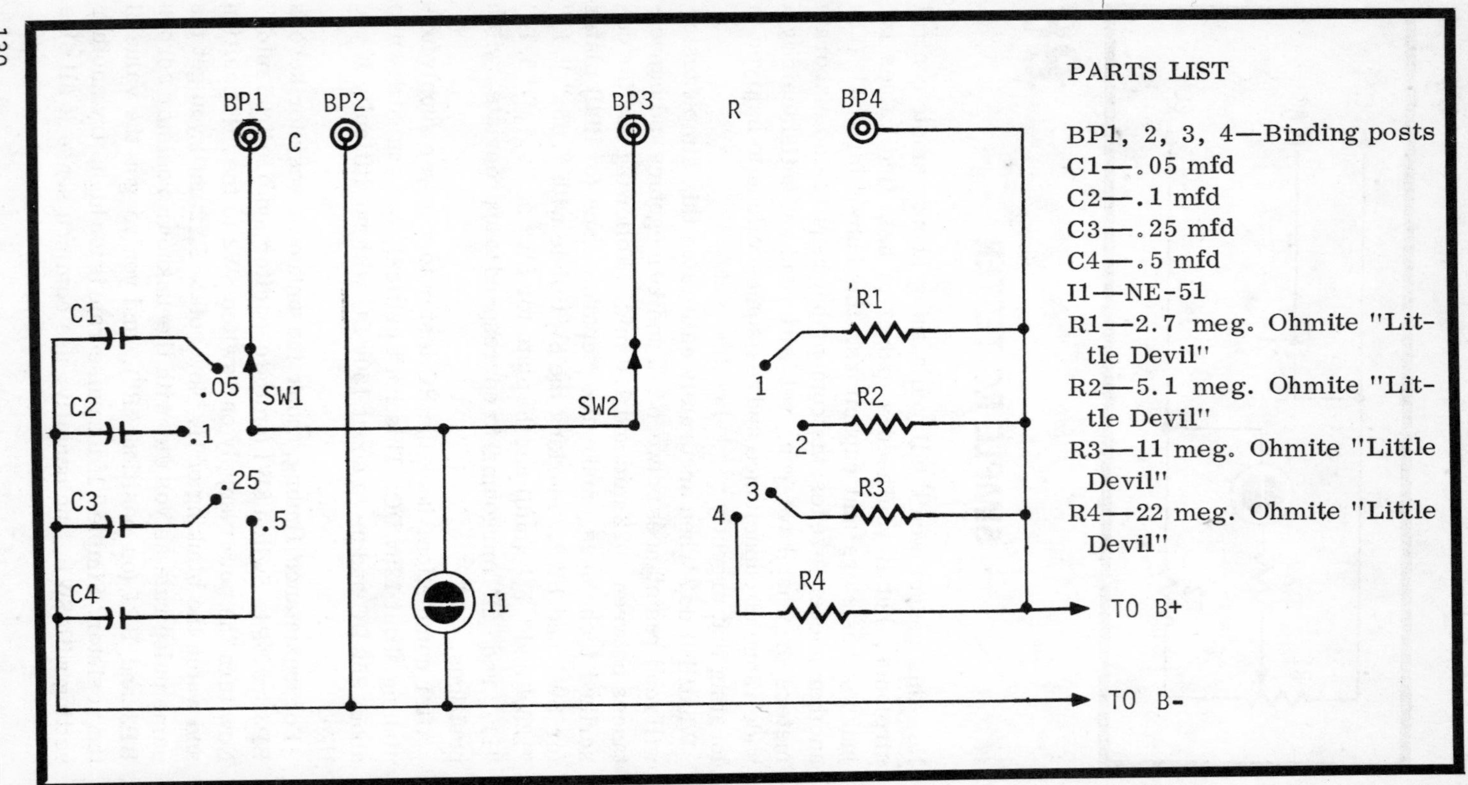
BP1
BP2
BP3
BP4
C
R
SW1
SW2
C1
C2
C3
C4
.05
.1
.25
.5
I1
R1
R2
R3
R4
1
2
3
4
TO B+
TO B-
PARTS LIST
BP1, 2, 3, 4—Binding posts
C1—.05 mfd
C2—.1 mfd
C3—.25 mfd
C4—.5 mfd
I1—NE-51
R1—2.7 meg. Ohmite "Little Devil"
R2—5.1 meg. Ohmite "Little Devil"
R3—11 meg. Ohmite "Little Devil"
R4—22 meg. Ohmite "Little Devil"

fails. If the pulsing rate is about twice what you're getting at the "2.7 meg" setting, chances are the resistor's a 1.25-meg job. If the pulse rate seems to be something between, say, the 11 and 22 meg resistors, chances are it's on the order of 15 megs in value.

For capacitors, hook the unknown value across BP1 and BP2. Set SW2 to the 2.7-meg position and notice the pulse rate at which I1 blinks while SW1 is in the "C" position. Now switch to the calibrated positions from .05 to .5 mfd, respectively, and match the pulse rate using the techniques just described. When you attain the same blinking rate, you've nailed down that previously-unknown capacitor's value. If you have trouble matching the blinking, try another position (not "R") on SW2.

HI-FI NOTCH FILTER

Here's a filter that is technically not a filter at all but rather a "nulling" device. It's main purpose in life is to provide you with a cheap way of filtering (or nulling) out unwanted AC hum that invariably is introduced in even the best hi-fi sets. To

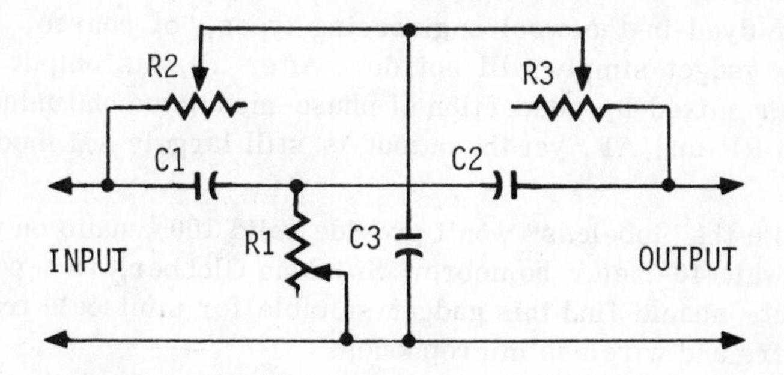

PARTS LIST

C1, 2, 3—.0047 mfd

R1—500K pot. Ohmite CMU 5041

R2, 3—2 meg pots. Ohmite CMU 2052

be pure about it, the filter doesn't have to be used strictly on hi-fi's either! You can build two if you have a stereo, or use a single filter in just about any speaker you have. This includes a TV set with an annoying hum or your Clegg Interceptor VHF Receiver. It requires no housing to speak of, and can be breadboarded or simply built into a plastic parts box.

In operation, you simply adjust potentiometers R1, R2, and R3 for virtual hum elimination. If you'd rather, you could substitute fixed values for the pots, but you'll lose the versatility and always wonder if you really got all that AC hum out of the system. (Rl can be 187.5K and R2 and R3 about 750K.)

90

TUBELESS/TRANSISTORLESS AM MODULATOR

Well, you're probably one of those chaps that said it couldn't be done. Nonbelievers please check the accompanying schematic diagram.

For dyed-in-the-wool engineering types, of course, this little gadget simply will not do. After all, its output is a rather mixed-up concoction of phase-modulated and unmodulated RF and AF, yet the output is still largely AM modulation.

While the "tubeless" won't provide solid 100% audio on your 750-watt 40-meter homebrew Swishing Clobber, flea-power addicts should find this gadget suitable for miniscule transmitters and wireless microphones.

Take care to keep the modulator shielded; about the only precaution necessary. A standard Minibox is just the ticket.

Incidentally, this modulator is great for tinkering around with in the lab. Ever have a hankering to put your Heathkit grid-dip meter on 144 MHz and give out with a CQ?

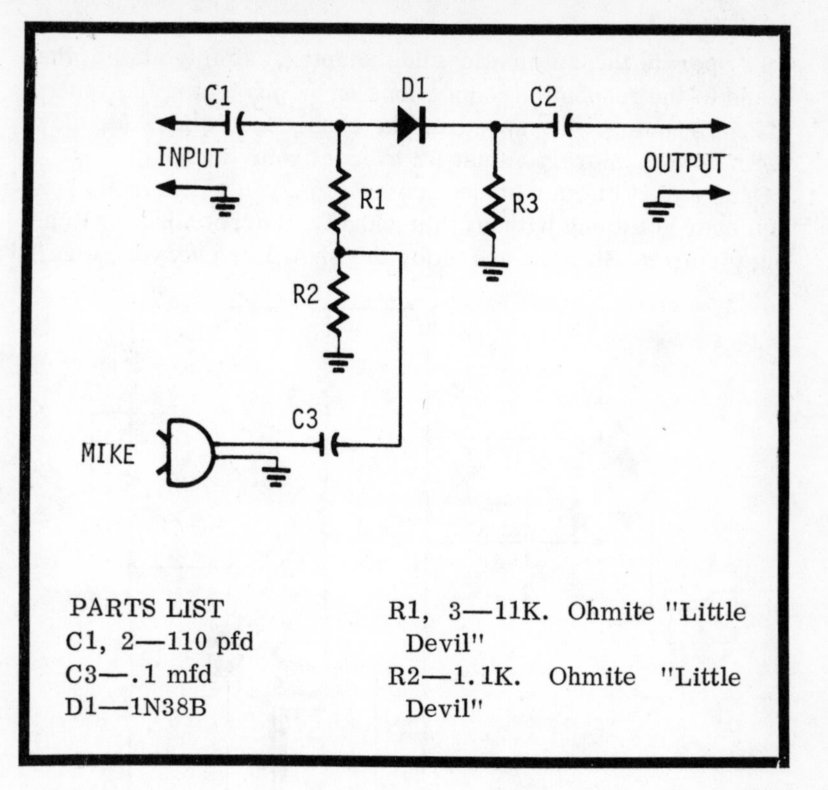

PARTS LIST
C1, 2—110 pfd
C3—.1 mfd
D1—1N38B
R1, 3—11K. Ohmite "Little Devil"
R2—1.1K. Ohmite "Little Devil"

91

AUTOMATIC PHOTO SLIDE PROJECTOR

In case you don't feel like laying out another $149.95 for one of those sleek, new automatic slide projectors, here's a project that employs no tubes or transistors, yet will automate any projector using a pushbutton advance control (as opposed to manual).

Aside from normal construction techniques employed when dealing with 117v AC, the only critical item is to insure that I1 is adequately close to photocell PC1 to trigger the relay. Light must fall on PC1 if it is to work. Obviously, however, this circuit should be housed completely so that stray beams of light don't fall on the photocell from any other source.

To operate the automatic slide adapter, simply attach the leads to the pushbutton connections on the projector and you're off and running. To speed up the changes of slides (or slow them down), merely adjust R2 to meet your fancy.

Once a part of your projection system, you will wonder how you ever got along without this gadget. Forget about switching pictures. Sit back and enjoy the show like everyone else!

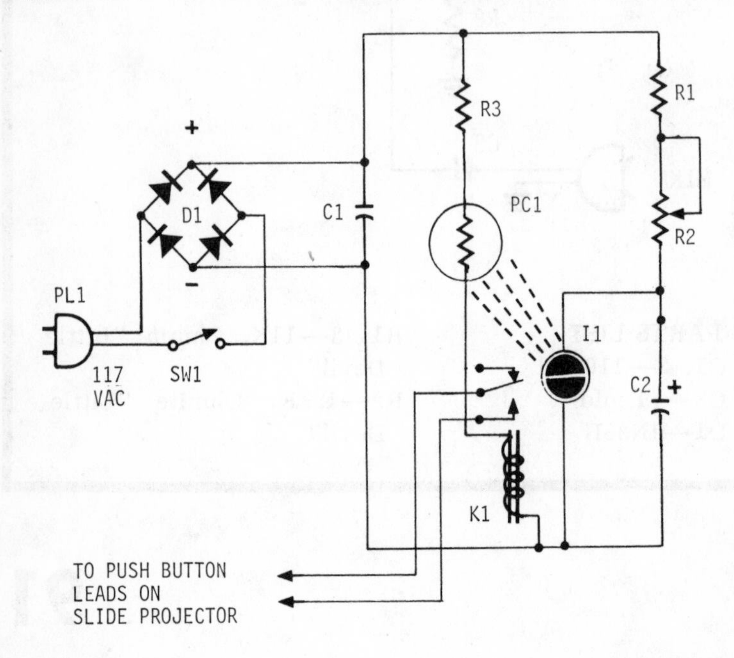

PARTS LIST

C1—10 mfd, 150 WVDC electrolytic
C2—20 mfd, 150 WVDC electrolytic
D1—Full-wave bridge rectifier. Mallory MTC-2314
I1—NE-2
K1—10K relay. Allied #75 U 774, Type LB-5
PC1—Photocell. Allied 7 U 565, Type LDR-C1
PL1—AC plug. Amphenol 61-M11
R1—18 meg. Ohmite "Little Devil"
R2—5 meg pot. Ohmite CMU 5052
R3—5.1K. Ohmite "Little Devil"
SW1—SPST. Oak Type 175

92

HANDY TUBE TESTER

Here's an extremely inexpensive checker that you can put together inside of an hour. It will test tube-filament continuity on all standard tubes.

Essential parts are just a handy box upon which to mount components, three tube sockets, a neon bulb, one resistor, and an AC line cord. What could be simpler?

Choose sockets that are fairly representative of the tubes you'll be checking; if you stick to the schematic and use one each of the following: 8-pin octal, 9-pin, and 7-pin you're sure to catch nearly all types.

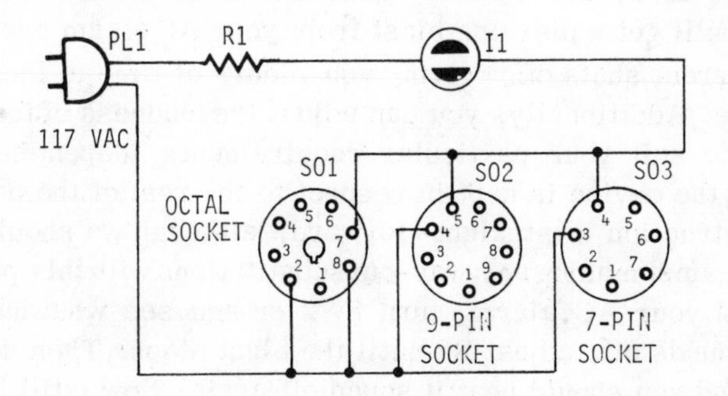

PARTS LIST
I1—NE-2
PL1–AC plug. Amphenol 61-M11
R1—110K. Ohmite "Little Devil"
SO1—Octal socket. Amphenol 88-8X
SO2—9-pin socket. Amphenol 59-409
SO3—7-pin socket. Amphenol 147-500

To test your tester, insert a tube that is known to be good into the appropriate tube socket. The NE-2 bulb should glow. If it doesn't, recheck your wiring. Now remove the tube. The glow should go out. Operation with unknown tubes is identical. If they're good, the bulb will light. If not, chuck them out.

93

AC LINE ALARM

Ever have your electricity go off in the middle of the night and not find out about it until early the next morning? Here you are, rushing around the house throwing out spoiled food in the refrigerator, checking which appliances were left on, what caused the problem, replacing fuses, and resetting all the blasted clocks, while all the time wondering if you are going to get to the salt mine on schedule.

Never fear, our Sonalert Line Alarm is here. From now on, you'll get a piercing blast from your AC alarm everytime the current shuts off, giving you plenty of time to locate the trouble. Additionally, you can adjust the loudness of the audio blast to suit your particular requirements (depending upon where the device is kept in respect to the rest of the house).

Construction is straight-forward, although we should caution against making any way-out substitutions with this project. To test your AC alarm, turn SW2 on and see what happens. If it sounds off, adjust R2 until the blast stops. Then depress SW1 and you should hear it sound off again. Now (still holding SW1 down) turn SW2 off. This will once again quiet the Sonalert.

If you want a louder sound, start with a larger battery supply. Use two 9-volt TR cells in series and drop a 5K variable resistor across the output to adjust the amount of voltage hitting the Sonalert. Starting all the way down (with virtually no voltage applied), adjust the potentiometer until you achieve the desired audio level. About 15 volts provides a healthy dose.

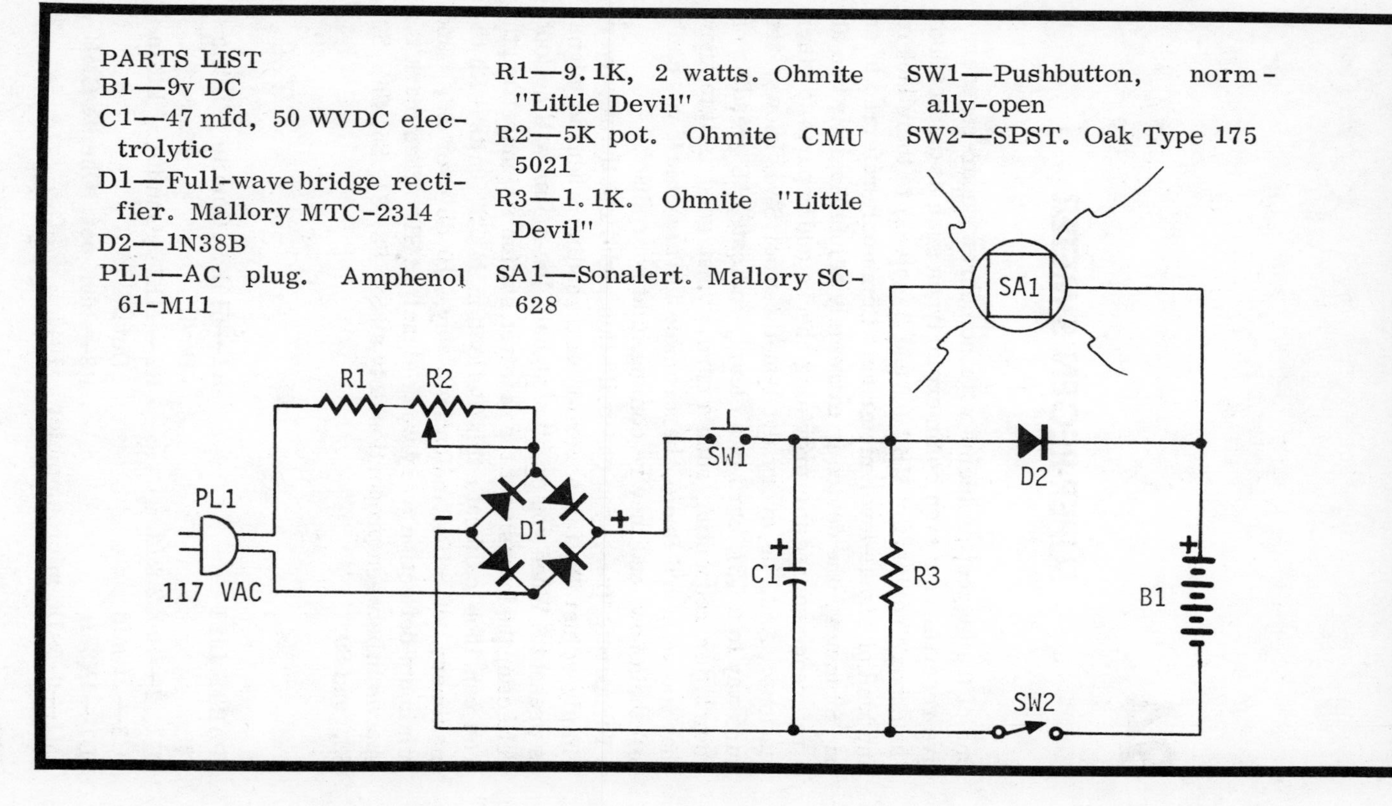
PARTS LIST
B1—9v DC
C1—47 mfd, 50 WVDC electrolytic
D1—Full-wave bridge rectifier. Mallory MTC-2314
D2—1N38B
PL1—AC plug. Amphenol 61-M11
R1—9.1K, 2 watts. Ohmite "Little Devil"
R2—5K pot. Ohmite CMU 5021
R3—1.1K. Ohmite "Little Devil"
SA1—Sonalert. Mallory SC-628
SW1—Pushbutton, normally-open
SW2—SPST. Oak Type 175
R1
R2
PL1
117 VAC
D1
SW1
C1
R3
SA1
D2
B1
SW2

94

SUPER-REGEN S METER

One of the biggest problems with homebrew super-regenerative receivers (or even commercial types such as the Heathkit "lunchbox" series for CB, 6, and 2 meters) is their lack of any kind of signal-level metering. Commonly referred to as an "S" meter, the device is extremely helpful in daily use as a reference source for providing checks and reports to other stations. And, as every dyed-in-the-wool SWL knows, the sure way to a QSL card is a clearly-indicated "S" reading on your letter requesting such a card. What good is it to know that you've heard Radio Abercrombie in Tasmania if you can't tell them how loud they're coming in at your shack?

To operate (yes, you can build this right into the receiver) simply adjust R3 for a zero meter reading when no signal is present. When you do hit a signal, the meter will deflect, reflecting the intensity of the received station's signal strength. You can then calibrate this deflection in "S" units with the maximum deflection indicating, say, 10 db over S9, and minimum deflection (very weak signal) at S1. Then mark the spaces inbetween proportionately as S2, S3, S4, S5, S6, S7, S8, and S9.

PARTS LIST

C1, 2—.0022 mfd
C3—.1 mfd
D1—1N38B
M1—0-50 DC microammeter
R1—510K. Ohmite "Little Devil"
R2—11K. Ohmite "Little Devil"
R3—100K pot. Ohmite CMU 1041

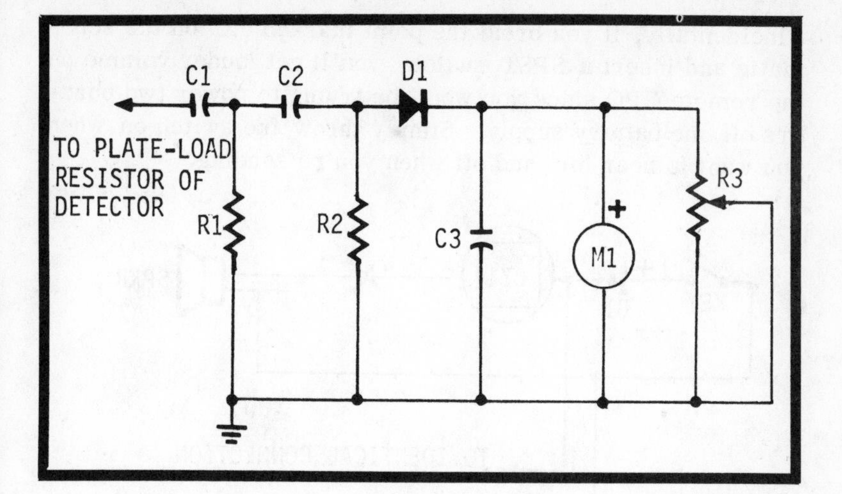

95

TUBELESS LOUDSPEAKER CPO

Here's a dandy gadget that at first glance may appear a bit crude, yet will surprise you in performance. It makes use of only four components: a key for sending CW, a 1.5-volt flashlight battery, a standard 1.5-volt high-frequency buzzer, and any handy PM speaker you happen to have laying around the junkbox.

Naturally, the resultant tone coming from the speaker is dependent upon the frequency to which the buzzer is set. For this reason it is well worth your while to invest 89¢ or so in an adjustable high-frequency buzzer, such as found in the catalogs of Allied Radio, Lafayette Radio, Olson Radio, etc.

If you want to add an embellishment, you can solder in a 1K potentiometer across the speaker leads. This will permit your adjusting the volume of the code practice oscillator.

By the way, this same scheme works well as one section of a wired two-way system. By running a pair of wires between your house and your buddy's, you'll have a dependable private CW communications link. It draws no current until somebody hits the key, so you don't have to worry about shutting the thing off.

Incidentally, if you break the point marked "X" on the schematic and insert a SPST switch, you'll get louder volume on the remote CPO since you won't be trying to power two buzzers off one battery supply. Simply throw the switch on when you want to hear him and off when you're sending.

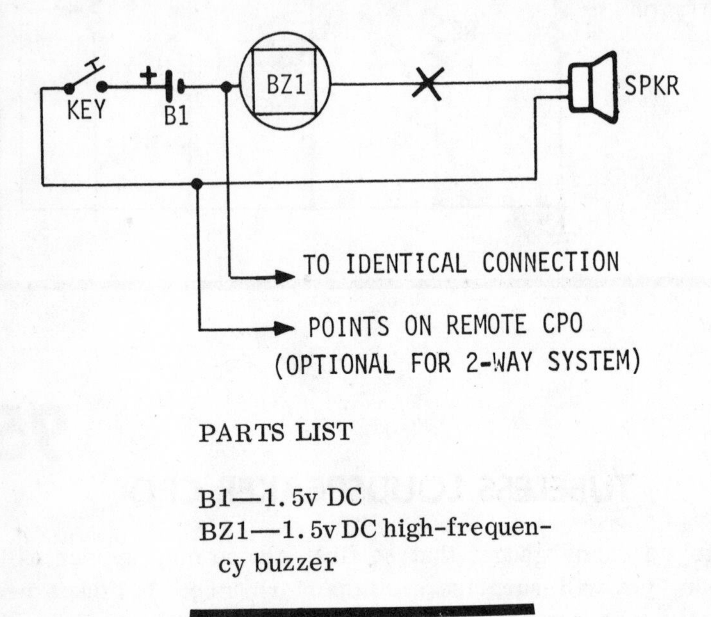

PARTS LIST

B1—1.5v DC
BZ1—1.5v DC high-frequency buzzer

96

LIGHT BULB POWER MEASURER

If you're a ham radio operator, you know all too well the value of determining the ratio of power input to power output of your transmitter's final stage. Yet aside from metering to learn power input, there really isn't a heck of a lot you can do to find out exactly what's getting up to the antenna without spending a week's wages on a sophisticated output power device.

This circuitry, however crude, will permit you to measure the output of the final for just the cost of a milliammeter, a 117v AC light bulb, an ohmmeter, and a few assorted capacitors.

The idea is to measure simultaneously the RF current flowing through a lightbulb and its DC resistance. Hence, all you do is apply the formula P equals I^2R (in watts, amperes, and ohms) and presto! You've got your power output! Once you get the hang of it, you'll wonder how you ever got along without this dandy gadget!

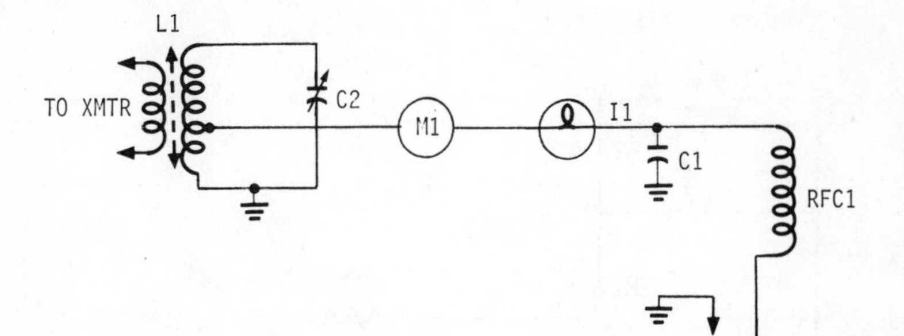

PARTS LIST

C1—.01 mfd
C2—Appropriate value to resonate with L1 at the transmit frequency
I1—100-watt light bulb
M1—0-500 DC milliammeter
L1—Appropriate tunable coil arrangement to resonate with C2 at the transmit frequency
RFC1—Appropriate RFC to protect the ohmmeter from burnout. Determine by the transmit frequency. (Example: Use Ohmite Z-27 at 27 MHz, Ohmite Z-50 at 50 MHz, Ohmite Z-144 at 144 MHz, etc.)

97

RF "SNIFFER" PROBE

Sure, you've seen lots of so-called sniffers, snoopers, detectors, probes, etc. But this one's completely different, both in principle and operation.

The most advantageous feature of this handy gadget is that it can detect extremely small quantities of RF. Hence, it is useful not only in conventional circuit troubleshooting applications, but also to find out if the oscillator's really oscillating, finding parasitics, even running down tiny low-power listening devices!

It is suggested that the probe be built into a tubular form, such as a modified fountain pen or the like for convenience's sake, although you can fashion this one any way you like and it'll still work. Simply adjust R3 for the sensitivity desired and you're in business.

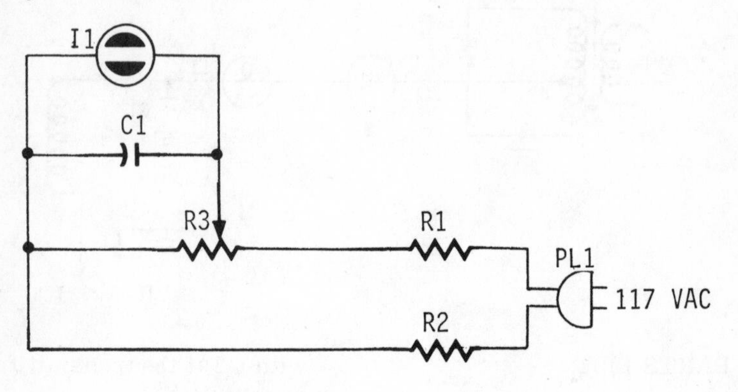

PARTS LIST

C1—.0022 mfd
I1—NE-2
PL1—AC plug. Amphenol 61-M11
R1, 2—1.5K. Ohmite "Little Devil"
R3—100K pot. Ohmite CMU 1041

98

CHEAP PUSH-TO-TALK RELAY

If you are a mobile nut who either uses a relatively high-power ham rig or a CB transceiver not originally intended for push-to-talk operation, you'll find this circuit extremely useful. Heavy-duty contacts are frequently not available in junkbox-variety relays, so we've simply "stolen" one from an automobile horn relay. You can buy the whole assembly for less than $1.00 from an auto supply store in your neighborhood.

Aside from the obvious economic savings this relay provides, it's also just the thing for your purpose, since it is well shielded from corrosion, dirt, etc., by a metal box and it is ready for immediate mounting on the firewall of the automobile.

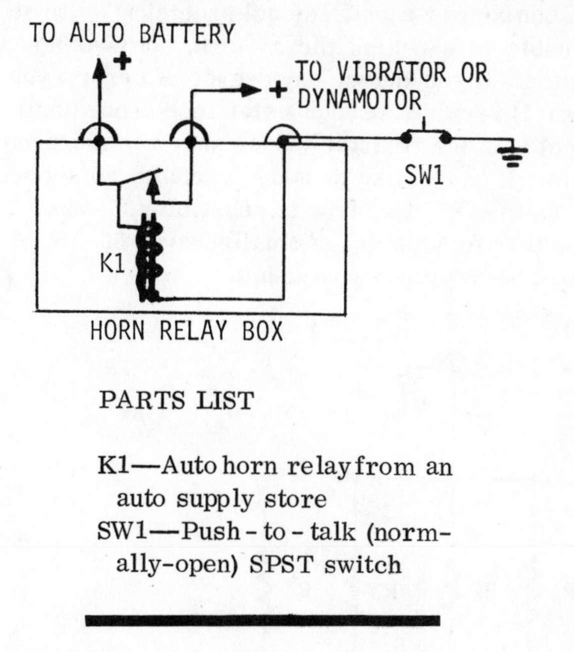

PARTS LIST

K1—Auto horn relay from an auto supply store
SW1—Push-to-talk (normally-open) SPST switch

99

DIRECT-READING CB POWER WATTMETER

Ever wonder how much power your Class D Citizens-Band transceiver or walkie-talkie is actually putting out? The easy-to-build direct-reading power wattmeter circuit shown in the accompanying schematic provides an inexpensive means for measuring outputs through four watts, yet it can be constructed with an absolute minimum of readily available junkbox variety components. The meter is a conventional 0-1 milliammeter, such as used for "S" meters and the like.

Keep all leads as short as possible and use a heat sink when soldering D1 into the circuit. You'll probably want to use a 5 x 3 x 2" Minibox to house your power wattmeter.

Either cut out the meter face drawn here (if it'll fit) or make a duplicate, pasting it over the old face with a slow-drying cement. You'll want to adjust the new meter scale to exact calibration (screwdriver adjustment on the meter front).

If you can borrow a professional precision wattmeter, it will be valuable in checking the accuracy of your new CB power wattmeter. If you detect any degree of error, you can elect to either 1) recalibrate your meter face accordingly or 2) experiment with the resistor values shown in the accompanying diagram. If you'd like to add a variable adjustment to your power wattmeter, feel free to substitute 35-ohm, 2-watt potentiometers for R5/R6. Actually, even 50- or 75-ohm pots will work when properly adjusted.

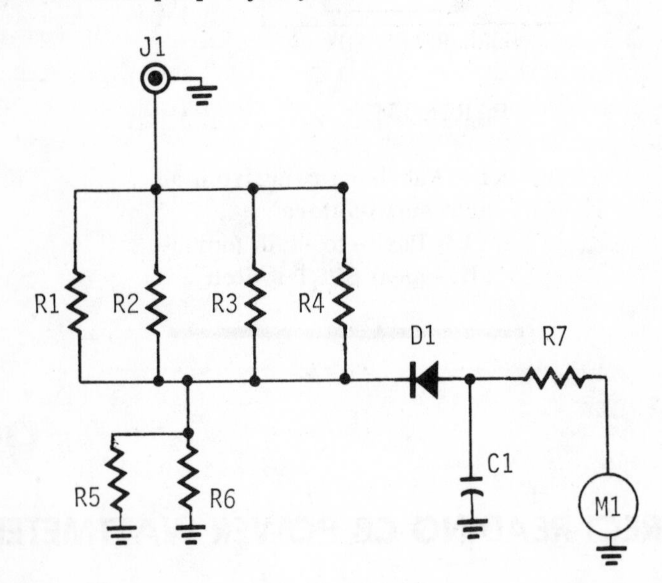

PARTS LIST

D1—1N38B

C1—270 pfd

J1—SO-239 coax receptacle. Amphenol 83-1RTY

M1—Small 0-1 ma "S" meter

R1, 2, 3, 4—160 ohms, 1 watt. Ohmite "Little Devil"

R5, 6—33 ohms, 1 watt. Ohmite "Little Devil"

R7—6.8K. Ohmite "Little Devil"

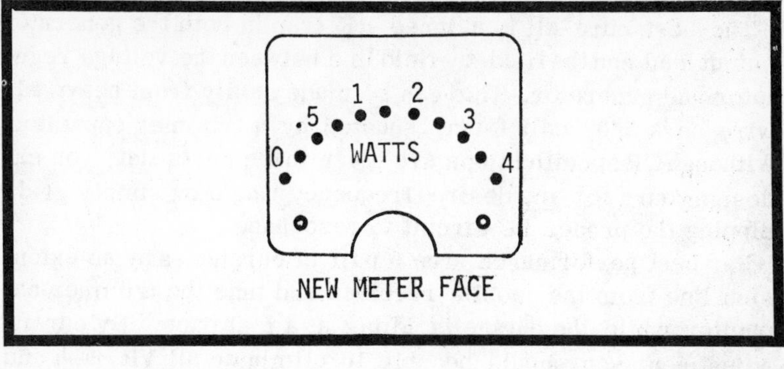

100

VOLTAGE REGULATOR HASH TRAP

One source of interference to an automobile receiver (any type CB, ham, car radio, etc.) which may cause a terrific "hash" is the normal sparking of the voltage regulator points. Although most mobile'ers prefer to simply employ a .15-mfd capacitor, manufacturers of these units warn against the use of such capacitor-to-ground arrangements which, if used, burn the points in no time at all!

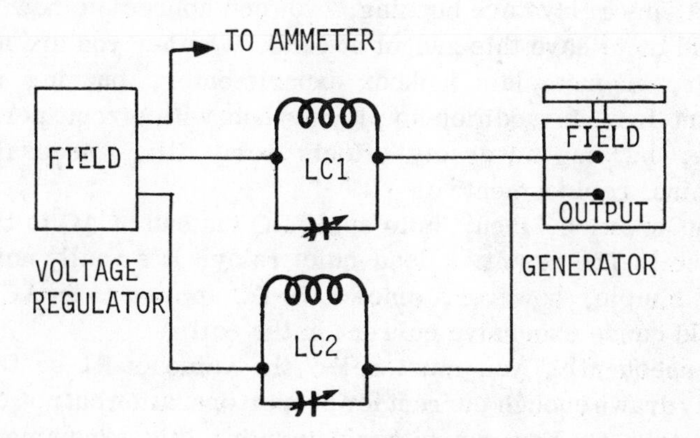

PARTS LIST

LC1, 2—Tuned circuit at the frequency of operation. (Example: For CB, use 12 turns of #14 wire closewound on a 5/8" diameter form, shunted with a 15-pfd trimmer capacitor adjusted for minimum noise at the receiver). Use a grid-dip meter.

The best cure-all is a tuned RF trap in both the generator output lead and the field-to-field line between the voltage regulator and generator. This can be made easily from heavy #14 wire on a 5/8" coil form, shunted by a trimmer capacitor. Although CB specifications are given in the parts list, you can design yours for any desired frequency range by simply grid-dipping the proper LC circuit to resonance.

For best performance, run a pair of earphones by an extension line from the mobile receiver and tune the trimmer capacitor while the engine is idling at a fast pace. By careful adjustment you should be able to eliminate all VR hash and quite a bit of generator "whine" too!

101

UNIVERSAL AC RELAY SILENCER

If you've ever had to transmit something like "Standby a moment, my relays are buzzing," you can appreciate how nice it would be to have this gadget at hand. Whether you are a ham, CBer, or just plain junkbox experimenter, buzzing relays are no fun. In addition to playing hob with circuit performance, buzzing adversely affects relay life, necessitating eventual replacement.

The solution, then, is to apply DC instead of AC to the relay coil. The result: a dead-quiet relay. It's really not quite this simple, however, since 117v DC applied to an AC relay would cause excessive current in the coil.

Consequently, you must select the value of R1 so that the relay draws enough current for proper operation but not enough to overheat. You can either determine this experimentally, starting, say, with a value of about 1K (at 20 watts) and moving up, or by simply installing a higher-value potentiometer as shown in the diagram and then by trial-and-error determine the proper permanent setting. In any case, the result will be well worth the effort, and never again will you be plagued with buzzing relays!

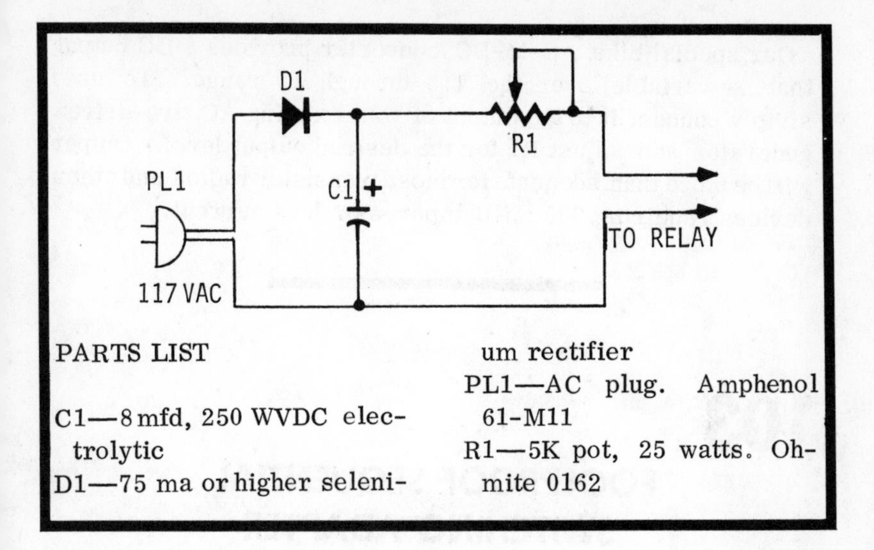

PARTS LIST

C1—8 mfd, 250 WVDC electrolytic
D1—75 ma or higher selenium rectifier
PL1—AC plug. Amphenol 61-M11
R1—5K pot, 25 watts. Ohmite 0162

102

BIKE GENERATOR AC-TO-DC CONVERTER

Have you or one of your youngsters ever wished for a way to convert the AC current from a bicycle generator to usable direct current? Well, as demonstrated in the accompanying diagram, there is a way.

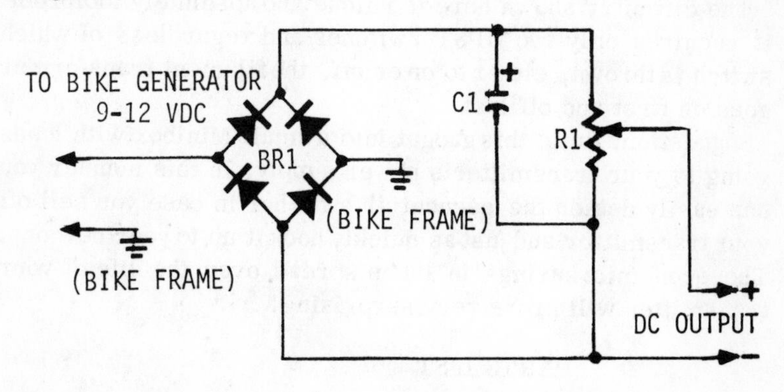

PARTS LIST
BR1—Full-wave bridge rectifier. Mallory 1B12B
C1—2000 mfd, 15 WVDC electrolytic
R1—35-ohm wirewound potentiometer. Ohmite Type 0109

Our special bike AC-to-DC converter provides a DC output that is variable over the 1.5 through 6v range. To use, simply connect it to the output of your existing AC tire-driven generator and adjust R1 for the desired output level. Output will be more than adequate for most transistor radios and other devices requiring 900 milliamperes or less current.

103

FOOLPROOF SEQUENTIAL SWITCHING ADAPTER

Anyone who has ever "homebrewed" a lot of radio equipment knows well the old story of "filaments on first and off last." While this may not seem that important at first glance, try firing up a ham radio rig the other way around and you may be very sorry you did! Indeed, several major TV manufacturers are at this writing considering an automatic sequential switching system to be incorporated into new sets to prolong tube life, especially in sweep circuits.

The circuitry shown here is unique and absolutely foolproof. It requires only two DPST switches and regardless of which switch is thrown, either to on or off, the filament transformer goes on first and off last.

Suggestion: Build this gadget into a small Minibox with leads going to your transmitter's power supply. In this manner you can easily detach the sequential switcher in case you sell off your transmitter and just as quickly hook it up to your new one. The economic savings in tubes spread over the life of your transmitter will prove very surprising!

PARTS LIST

PL1—AC plug. Amphenol 61-M11

SW1, 2—DPST. Household-type AC switches

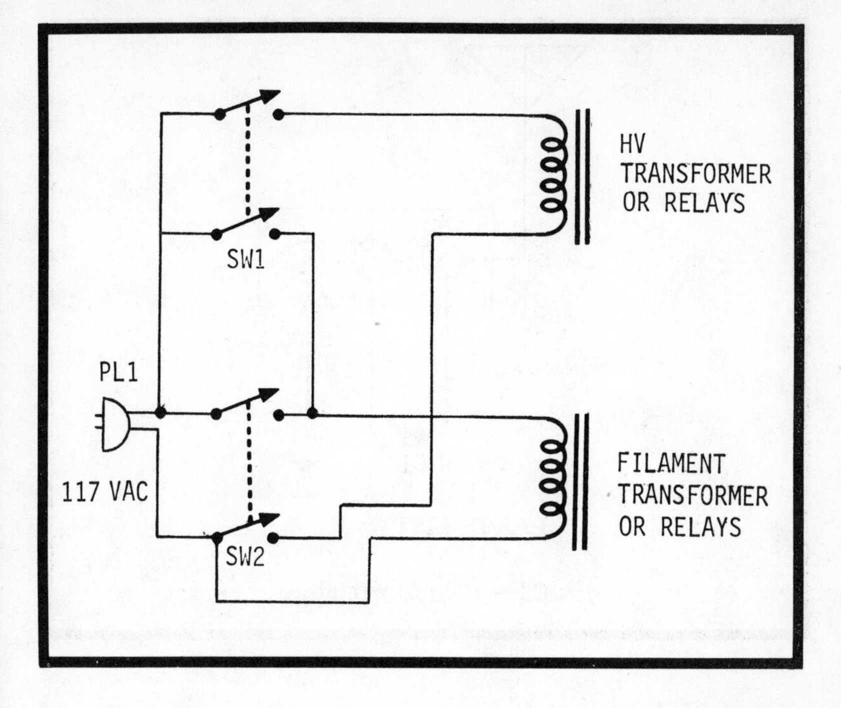

104

SHORTWAVE TRIMMER BOOSTER FOR VINTAGE RECEIVERS

While almost too simple to appear effective, this system can do wonders for your aging, low-priced shortwave receiver. Essentially, it acts as a scaled-down preselector that "hops" up the signal before it enters the receiver, resulting in amazing performance.

If you listen on 7, 14, 21 and/or 28 MHz (yes, you can modify this for the CB frequencies), this technique shouldn't be overlooked. Use a low-voltage variable capacitor of 100 to 140 pfd connected as shown in the diagram.

Most receivers have room inside where the trimmer can be installed, although it will not adversely affect performance if an outboard installation is made. In operation, merely tune the capacitor for maximum signal as shown by the "S" meter.

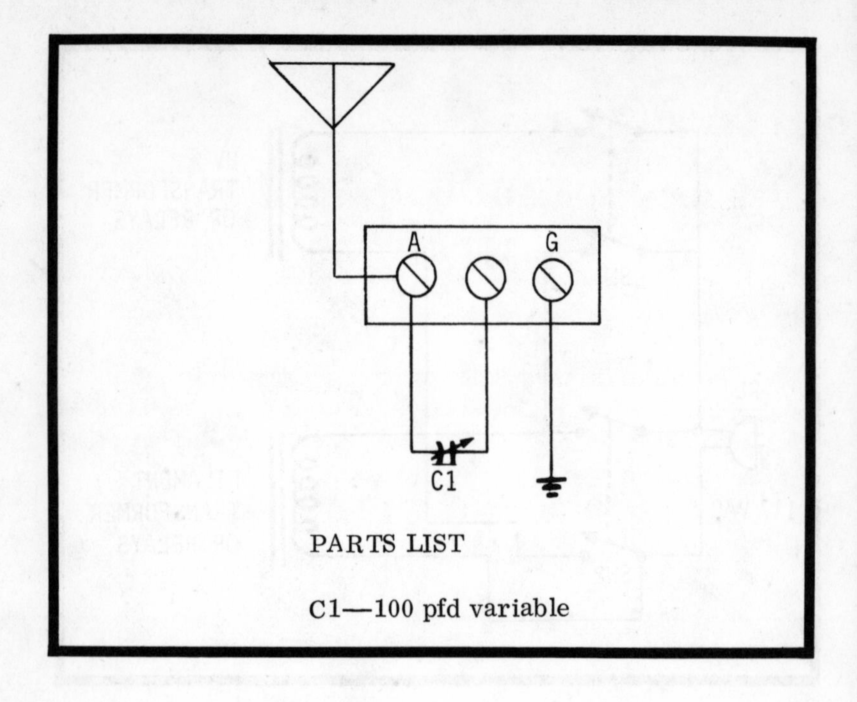
A
G
C1
PARTS LIST
C1—100 pfd variable

SUBSTITUTION GUIDE

Here are standard substitutions for most of the diodes and neon bulbs used in this book. All substitutes have not been tried in the circuits and some slight variance in circuit operation is possible and should be expected. Some substitutions will, in fact, improve overall operation. Where no substitute is shown, no practical one was found.

Diodes	Substitute
1N34	1N34A, 1N38B
1N34A	1N34, 1N38B
1N38B	1N38, 1N38A
1N60	1N34A, 1N54A, 1N64, 1N64A, 1N295
1N2070	1N1695, 1N1763, 40H, GE504, 1N2863, F4

Neon Bulbs	
NE-2	NE-51
NE-2A	NE-51
NE-16	NE-45, NE-48
NE-30	NE-32, NE-34
NE-48	NE-57, NE-21, NE-17, NE-7
NE-51	NE-2, NE-2A
NE-83	NE-23

PARTS SUPPLIERS

Allied Radio Corp., 100 N. Western Ave., Chicago, Ill. 60680.

ARC Sales, P.O. Box 12, Worthington, Ohio 43085.

Arcturus Electronics Co., 502 22nd St., Union City, N.J. 07087.

Arrow Sales, Inc., 2534 S. Michigan Ave., Chicago, Ill. 60616.

Barry Electronics, 512 Broadway, New York, N.Y.

BC Electronics, 2333 S. Michigan Ave., Chicago, Ill. 60616.

Burstein-Applebee Co., 1012 McGee St., Kansas City, Mo.

Columbia Electronics, 4365 West Pico Blvd., Los Angeles, Calif. 90019.

Edmund Scientific Co., Barrington, N.J. 08007.

Fair Radio Sales, P.O. Box 1105, Lima, Ohio 45802.

FM Sales Co., 1100 Fremont St., Roxbury 20, Mass.

General Surplus Sales, 10 Alice St., Binghamton, N.Y.

G & G Radio Supply Co., 77 Leonard St., New York, N.Y. 10013.

Jeff-Tronics, 4791 Memphis Ave., Cleveland 9, Ohio.

J.J. Glass Electronics, 1624 South Main St., Los Angeles, Calif.

Joe Palmer, P.O. Box 6188, Sacramento, Calif.

Lafayette Radio Electronics, 111 Jericho Turnpike, Syosset, L.I., N.Y. 11891.

McGee Radio Co., 1901 McGee St., Kansas City, Mo. 65108

Mendelson Electronics Co., 516 Linden Ave., Dayton, Ohio 45403.

Meshna, 19 Allerton St., Lynn, Mass. 01904.

Olson Electronics, Inc., 260 South Forge St., Akron, Ohio 44308.

Poly Paks, P.O. Box 942, Lynnfield, Mass.

Radio Shack Corp., 730 Commonwealth Ave., Boston, Mass. 02117.

R.E. Goodheart Co., Inc., Box 1220, Beverly Hills, Calif. 90213.

Space Electronics Co., 218 West Tremont Ave., Bronx, N.Y.

Solid State Sales, P.O. Box 74, Somerville, Mass. 02143.

TAB, 111 Liberty St., New York, N.Y. 10006.

Telemarine Communications, 142 West Broadway, New York, N.Y. 10013.

World Radio Laboratories, 3415 West Broadway, Council Bluffs, Iowa 51501.

Index

A

B

C

D

E

F

G

P

R

S

T

V

W